THE FRENC

CW01664331

'. . . has the adva
veteran, an ex-offi
and straightforwa... and rich in unpublished material, for example, the Legion's final days in Sidi bel Abbes after Algerian independence.'

'. . . noteworthy for its authentic anecdotes . . .'

'. . . a history which is also a tribute.'

Titles in the Corps d'Elite series:

THE FRENCH FOREIGN LEGION by Erwan Bergot
THE SAMURAI by Jean Mabire & Yves Breheret

to follow:

THE AFRIKA KORPS by Erwan Bergot
THE COSSACKS by Yves Breheret

THE FRENCH FOREIGN LEGION

Erwan Bergot

Translated from the French by
Richard Barry

TATTOO

A TATTOO BOOK

published by

WYNDHAM PUBLICATIONS

A Tattoo Book
Published in 1976
by Wyndham Publications Ltd.
A Howard & Wyndham Company
123 King Street, London W6 9JG

Published in France by Balland, under the title *La Légion*, 1972
© Balland, Paris, 1972
First published in Great Britain by
Allan Wingate (Publishers) Ltd., 1975
Translation copyright © by
Allan Wingate (Publishers) Ltd., 1975

The author and the French publisher acknowledge the help given them by the Historical Department of the Foreign Legion in providing historical documents and illustrative material.

Printed in Great Britain by
Richard Clay (The Chaucer Press) Ltd., Bungay, Suffolk

ISBN 0 427 00011 4

This book is sold subject to the condition that it shall not, by way of trade or otherwise, be lent, re-sold, hired out or otherwise circulated without the publisher's prior consent in any form of binding or cover other than that in which it is published and without a similar condition including this condition being imposed on the subsequent purchaser.

Contents

Appendices

Preface

ALTHOUGH ITS military value has never been questioned, the French Foreign Legion has not enjoyed a universally good reputation. Many serious works of scholarship have been devoted to it, but, in the minds of a considerable proportion of the public, they have been overshadowed by preconceived ideas, clichés or even the prejudices resulting from the picture painted by the mass media.

The prime virtue of this book, therefore, undoubtedly is that it enables the reader to acquire a far more accurate idea of the Legion. Of course, not all legionaries are angels, but the majority of them are by no means adventurers, drop-outs or freebooters. They are men who have become soldiers, often in the finest sense of the word. The idealists among them were in fact legion and, when all is said and done, far more important than the individual failings is the spirit of the Legion, its traditions and its fighting qualities which have made it a *corps d'élite.*

Through the history of the Legion and the way in which the author has approached it, Erwan Bergot's book enables us to appreciate with greater objectivity what the Legion was, and what it still is today. In addition to this, however, one conclusion, and in my view a far more important one, will inevitably strike the reader: after certain trial-and-error experiments, in no other military formation has there been such thorough-going fusion of soldiers from such differing national origins, although it is true that almost all came from the same great family of European nations. During their long history these nations have been involved in crusades, coalitions and alliances; they have formed international brigades and 'grand armies'. But there has never been another Legion animated by a similar *esprit de corps* or inspired by a similar tradition unless one goes back to the times of the ancient orders of chivalry.

The legionaries' flag carries the words '*Legio, patria nostra*'. In a sense the Legion in itself constitutes a sort of country and it has even been said, in my view wrongly, that basically a *corps d'élite* has no country other than itself. Could it not be, however, that throughout all the history which it has made for itself in every part of the world, one of the hallmarks of the Legion has been this sense of affiliation with Europe? It is a sentiment which springs not only from its origins but from its tradition as pioneer, builder and colonist, thereby connecting it directly with those other legions which, in the great days of the Empire, Rome despatched to its most distant marches, from the confines of Mauretania to those of Bactria.

The essentially European character of the Legion should be remembered and taken into account now that Europeans can see the day approaching when they will really have to organise their defence in concert. The facts that the Legion is steeped in French military tradition and that its official language is French make no difference in this case; those who drafted the European Defence Community treaty – genuine Europeans if ever there were any – had envisaged French being the main language of the European defence forces. Similarly I have myself frequently advocated French as the working language of Europe.

Since its background is European, therefore, should not and ought not the Foreign Legion to come to Europe or rather return to it? Instead of being 'foreign', could it not officially become that which in fact it is, in other words European? Could it not find its niche in a new community defence organisation, the European equivalent of what the Marine Corps is for the United States of America?

The idea has already been broached early in the 1960s by

the 'European Study and Action Centre'. It is an idea which should be revived as Europe expands, organises itself and reaches the stage when it must seriously consider the great problem of its defence. In building its future Europe cannot afford to overlook any of the factors of cohesion and strength or any of those living traditions which it can preserve – and should maintain – as a heritage from its great past.

So this book, which fits so well into a series devoted to *corps d'élite*, should not be read merely as an account of the Foreign Legion's past and a description of its present. If it leads people – and I hope they will be many – to reflect on the potentialities of a European expansion of this illustrious military formation, the book may make a concrete contribution to preparation for the future defence organisation of a United Europe.

OTTO OF HABSBURG

Foreword

WRITING A history of the Foreign Legion is a risky thing to do. A monument does not make a story and the author requires a degree either of ignorance or presumption to think that in a few hundred pages he can do justice to a century and a half of battles, campaigns, gallantry and courage or give in a book a true picture of the life of the three hundred thousand men who have worn the Legion's cap-badge since 1831.

Others have already tried and the plethora of literature creates its own difficulty; the subject is so rich, has been dealt with so often that it seems to bear down on the author with a crushing weight. The Foreign Legion is a challenge. The author who takes up the challenge is in danger of becoming bogged down in the mass of eyewitness accounts, of stories and of anecdotes; all the time he is paralysed by his main preoccupation – to do justice, to bear witness and to serve the truth. What to say differently? What to say more? How to say better what has already been said?

Nevertheless the story should be told; papers a century old and reports which no one today has either the time or opportunity to read should be brought to light. This is the sole justification for an author to write about the Legion in 1972.

Then there arises another problem, that of selection. It is not possible to tell the whole story and so, from the outset, the question presents itself: why this battle and not another one? Why this legionary and not his fellow? Why this regiment and not its neighbour? No more than the battle itself, courage, death and glory are not the preserve of a few. The Legion is a whole.

Selection therefore, however arbitrary it may seem, looks for the symbol – in telling the story of the battle of a few, one is thinking of the courage of all. In this way each man, whether hero or forgotten, whether well known or anonymous, receives his due measure of the truth.

In this book the name 'Camerone' frequently recurs. For the Legion it is more than an example; it is an attitude of mind. Naturally every legionary is proud of this battle conducted at odds of fifty to one. He is proud of it, as he is proud of all those other 'Camerones' which took place elsewhere, at other times and under other skies. Fifty to one seems to be the usual odds for the Legion's battles.

In addition to the feat of arms, however, there is a day-to-day reality – the spirit of Camerone. It emerges as a series of simple notions and ends as the professional virtuosity of the regular soldier. For the legionary is a professional with all that that implies in conscientiousness, seriousness and sense of efficiency. Everything else is marginal.

When one remembers that this virtuosity will lead a man to die for an oath that he has sworn, that his professional conscientiousness compels him to fight a good fight when the interests of the country so require, it is a far cry to the legend, the myth or the stereotyped picture of the 'lonely desperado' searching for 'oblivion' under the *képi* of the faceless legionary.

One old sergeant said to me: 'Write what you like about the Legion but at least say that we were soldiers – genuine soldiers. Of course we were tough and of course we were difficult. But one has to be tough in order not to crack when the less "tough" collapse. One has to be difficult because war is not a simple matter. One gets killed even if one has no wish to be. Death is there and the fact that one has accepted it as a possible risk is no reason to court it at any price. A legionary dead is a legionary who will not be there for the forthcoming counter-attack.'

Admittedly today there is no more war and the Legion has other things to do. But the spirit is still there and whenever they build something, be it at Aubagne, at Djibouti or in the Pacific, the hallmark of the legionaries is still to be found: the sense of something 'completed', the conscientiousness of the professional.

Forty years ago Colonel de Corta, one of the great figures of the Legion in Africa, gave the following description of his men and it is still as valid today as it was then.[1]

> Legionaries are masons, navvies, electricians, plumbers, carpenters, artificers or mechanics – anything you like. They know how to paint, sew, bind, rough-cast a wall, decorate a room, put up a roof or build a road as well as they can attack or listen to Schumann or Strauss when the band is playing by the sea in the evening. They are professionals in the broadest sense of the word.
>
> We are not dealing with schoolboys here. Our soldiers are men who know about life and whom one cannot lead up the garden path. They have all already worked, struggled and travelled; their professions are as different as their nationalities. That is why, with them, one can do anything.
>
> How old are they? They do not have to tell us. We only know that they have done ten, fifteen or twenty years in the Legion.
>
> One day, perhaps rather late, they realised that they were cut out to be soldiers; today that is a difficult career; Europe now only has militia; but the Legion remains and it offers all these men who have no purpose, no security and are divorced by years of silence from their families if they have any, a great big family where the family spirit reaches its height. All officers of the Legion know that they must provide a substitute for the ties which these men lack, that wherever they go, they must create a home in which their men will find moral security by living cheek by jowl, finally that they must provide that sort of happiness which comes from the action and violent effort which these men

1. Quoted by Jean des Vallières.

have come to seek under some forgotten sky. What we ask of them would be inconceivable in an ordinary unit. But they accept it because they know that in anything which concerns them personally we will invariably take the responsibility for understanding them, helping them and defending them up to the hilt, even if this, our first duty, puts our career at stake.

Obviously one can hardly conceive of them in a garrison town like Melun; things are not always easy when they are off duty in a town. But you should see them in the field or in a scrap, when at last they are living the life they want; they have extraordinary dash and a sense of iron discipline; they also have their splendid uniform and, throughout the ages, this has mattered to great armies. Troops are like women in some ways; those who have no sense of showmanship are not worth much.

Often they act on impulse; men always enlist in the Legion somewhat on impulse and that postulates character, a determination to fall on one's feet which not everyone possesses. During a war many of your friends will die bravely; do you know how many who are able, one fine day, to rub out everything, not in order to die but simply by saying: 'We'll make a fresh start'? As for the 'truth', which is no business of ours but which we sometimes guess, it is of no interest to the newspapers; it is too simple, too ordinary, too 'old hat'. But, believe you me, there are always resolute, high-spirited young men who feel the sense of adventure and the need to use their strength and who find their calling in the ancient profession of the soldier, that of builder.

This is the spirit of Camerone. This is what the author has tried, very modestly, to bring out.

ERWAN BERGOT
Paris, January 1972

Donne ta main, mon camarade
Tu n'as pas tellement changé
Lorsque le Premier Étranger
Vint t'accueillir pour la parade
Maintenant tu es allongé . . .

Donne ta main, mon camarade
Car le bon sang ne peut se renier
Venu de Berlin? De Belgrade?
Mais tu es mort sans rechigner
Pour remplacer, seul et sans grade,
Un Français qui n'est pas tombé.

Ton Drapeau, grâce à ta bravade
A recouvert le monde entier.
Tu n'as pas tellement changé
Notre Inconnu, seul et sans grade
Depuis la première parade
On n'est pas près de t'oublier
Donne ta main, mon camarade.

Give us your hand, my friend.
You have not changed much
Since the day when the 1st Foreign Legion
Had just welcomed you on parade.
Now you are laid out cold . . .

Give us your hand, my friend.
The good soldier remains true to himself
Whether he comes from Berlin or Belgrade.
You died without recrimination,
Alone and unsung, in place of
A Frenchman who did not have to die.

Thanks to your bravery your flag
Has flown all over the world.
You have not changed much,
Our Unknown Soldier, alone and unsung,
Since that first parade.
We are not going to forget you.
Give us your hand, my friend.

J. VALREY

Le Képi blanc [Monthly magazine of the Foreign Legion], January 1972

Loyalty and Honour

His name was Martin. It was not his real name, of course, but who cares in the Foreign Legion what a man is really called? There people are judged by their qualities as men and on the battlefield, not by their status in law. He could have been called Schmidt or Ruiz or Novak but that would have made no difference to his qualities as a soldier and a leader.

He was a sergeant-major and that after only six years' service, something fairly exceptional, but then again promotion in the Legion is a question of ability rather than seniority. In six years Martin's qualities had transformed him from a young recruit into one of the most bemedalled NCOs in his unit, the 2nd Battalion of the 3rd Foreign Legion Infantry Regiment [3e R.E.I. – Régiment Étranger d'Infanterie].

The year before, 1950, he had won the Military Medal at Langson. Having run out of decorations with which to reward his acts of valour, the Commander-in-Chief in Indo-China could now do nothing other than bestow on Sergeant-Major Martin, Émile, the Cross of Chevalier of the Legion of Honour.

As was customary on such occasions, on this 12 July 1951 an outsize 'loving-cup' was organised in the NCOs' mess at Tien Yen as a baptismal ceremony for Sergeant-Major Martin's Legion of Honour or 'red ribbon'. The atmosphere was merry, relaxed and intimate. The men drank and sang 'Le Boudin' and other Foreign Legion marching songs. Everyone was happy, a little boisterous perhaps, and everyone talked loudly.

Everyone, that is, except the recipient of this honour, who seemed indifferent to the fame which he had won. He was drinking with the others, of course, but with a certain air of detachment. Clearly Sergeant-Major Martin, Émile, was thinking of other things.

In the heavy humid heat of the little post in Tonkin, scorched by the sun, infested with mosquitoes, surrounded by bamboos and red laterite boulders, he was thinking, no doubt, of that other 'loving-cup' for his first Legion of Honour, won in another place and under another name.

That had been seven years before during the Russian winter, somewhere near Stettin during the frightful 1944 retreat, with snow and shells falling. But Martin kept his memories of that period to himself. Then he had been Lieutenant de X, commanding a company of the French volunteer contingent on the Russian front. His men had been mostly descendants of Russian 1917 émigrés, bent more on avenging themselves on the Bolsheviks for their exile than on reconquering a country which they neither knew nor liked.

Subsequently he had been afraid that he might run across some of them as a result of his various postings in the Foreign Legion. One or two had in fact joined the Legion – Khrukov, for instance, who had enjoyed stringing up partisans near Smolensk, or Chorine son of a Kuban Cossack; he had been in another battalion – God knows under what name – and had been killed four years before near Phu Tong Hoa

Now, carefully and somewhat absent-mindedly stroking the brand-new red silk ribbon and the enamel of his cross, Martin was no doubt trying to recapture the exhilaration which he had felt at that time when awarded the highest French honour; the cross was oddly similar to one which he had won as a child for his first essay at school.

For an instant the picture came back, clear as a flashlight photograph. They had bivouacked in a dark, smoke-blackened church. Christmas was not far off – a real Christmas with snow and pine trees. His men had burnt the wood of the altar to keep warm and do their cooking – what did it matter? Then they had gathered in the choir stalls under the only section of roof still intact and had seen the cold distant stars shining through the twisted rafters. His platoon had huddled under their drab capes, stiff with mud and cold, helmet pulled down over the forehead, faces half hidden in their grey-green balaclavas.

They had found some alcohol and like automatons – or sponges – they had drunk this potent stuff with its strong sour taste of potato. It was cold, too cold, –40°C perhaps, and that explained everything; to warm up inside one had to drink, no matter what it was.

They had been neither merry nor sad. They were too fatalistic and too weary to have any feeling. So, sitting alone, Lieutenant de X, regular officer, 1938 graduate of St Cyr (the motto of his term had been 'Franco-British friendship'), smiled to himself from time to time with a macabre sense of humour about his Legion of Honour.

'Hi! Martin! Are you snoozing?'

The dream ended. Stroebel, the battalion warrant officer, was standing in front of him with a friendly smile.

'You're wanted in the personnel office.'

Martin's eyes widened. The personnel office? What on earth could the officer responsible for numbers and postings of NCOs and men in the various units want with him?

Martin went off puzzled, only afraid that his newly acquired Legion of Honour might end by doing him a bad turn like getting him posted to another battalion.

'Sergeant-Major Martin. Reporting for duty, sir.'

He had entered in the regulation manner, sweeping salute, chest thrown out, hand cut back to the side, heels clicked.

'Stand at ease, Martin,' the officer replied.

He pointed to a typewritten sheet of paper in front of him – an official telegram.

'I have just this moment had a telegram about you.'

Martin raised an eyebrow, his only indication of curiosity. He waited resignedly for the news. It is often this way in the Foreign Legion. Life consists of leaving one's buddies, to find others elsewhere. As a result of postings over the years he had been the rounds and had buddies everywhere – 'the great Foreign Legion family'.

'Martin, I am to tell you that you are entitled to an amnesty for anything done during the war of 1939–45. You see what I mean. There is nothing to prevent you reassuming your real identity.'

The officer left time for this news to sink in and then went on: 'As a special favour, in view of your record of service, you are authorised to resume the rank of officer if you wish. You would be a reserve officer on active service. Naturally you would rejoin in a unit of your choice in the regulars' [the term used for all non-Foreign Legion units].

Martin went pale. For barely a second he thought what this new life would be like – gold stripes on his tunic, other duties, another command . . . and his real name.

The lieutenant was about to get up and shake hands with this new fellow-officer whose bravery under fire was the admiration of all. His smile broadened. Leaning his left hand on the edge of the desk, he was about to hold out the right.

The sergeant-major took a pace to the rear. He drew himself up and threw out his chest. His heels clicked once more. He saluted and before turning about, as if afraid that he had not made his meaning clear, he said, 'Sergeant-Major Martin, Émile, six years' service, army number . . . reporting for duty, sir.'

The story is not of some feat of heroism; it is merely an anecdote arising out of war. But it illustrates a man's loyalty to the career of his choice. The history of the Foreign Legion is full of instances of this nature.

One day [General Bruneau says] I received a registered letter with a Viennese stamp on it. It came from the head of a well-known Austrian intelligence agency and ran as follows:

Sir, I should be grateful if you could let me know whether there is now in Saida [the Foreign Legion Training Centre] a young man of Austro-Hungarian nationality who is said to have enlisted in the Legion under the name of Justus Perth; enquiries from the 1st Regiment have been without result. You will understand why we are anxious to find him when I tell you in strict confidence that, as a result of an unforeseen occurrence, he has, without knowing it, become heir to a fortune of twelve million crowns.

Attached is a photograph of him taken when a student in Prague.

I had a look at the photograph. It showed a well-built young fellow, aged twenty to twenty-two, chubby-cheeked and with a short beard. He wore spectacles, so I could not see his eyes.

'Go and see the major,' I said to Ramus, my warrant officer. 'Ask him whether he has on his list a legionary named Justus Perth.'

A few minutes later the warrant officer returned shaking his head. 'No one of that name.'

'Very well. Collect all the Austrians in the platoon and we will check with this photograph.'

The warrant officer called the duty bugler and had the 'On parade' sounded. 'All present,' he announced.

'All right,' I said. 'Go down with Dhürmer, the German-speaking clerk, and go over all the men with him. Then bring along to me those who, in your opinion, look like this photograph.'

Ramus saluted and went down. From my window I could see him, together with the clerk Dhürmer, inspecting the legionaries one by one. Finally he selected two and brought them along to my office.

'Colonel,' Ramus said as he brought them in, 'there are only these two who look more or less like the photograph. They arrived with the last intake and do not speak French. However, they have both told Dhürmer that they are not Justus Perth.'

'Now then,' I said to the clerk, 'tell them what it's all about. That should loosen their tongues.'

'Here's what the colonel has just said,' Dhürmer explained. 'Justus Perth has just become heir to twelve million crowns. If one of you is the lucky dog in question, speak up.'

I went up to the first man :

'Is it you ?' – '*Nein.*'

'And you ?' – '*Nein.*'

The clerk then lost his temper. After telling them that they were complete asses, he repeated once more what they were expected to say. But they both shook their heads. They were not Justus Perth. I had to send them away and reply to the agency in Vienna that the man they were looking for was not in my unit.

Several years later – by which time I had forgotten the incident – I received a large official envelope, this time carrying the stamp of the Austrian Ministry of Foreign Affairs. The letter it contained drew a cry of astonishment from me. In diplomatic language it repeated the Vienna agency's request for information and insisted that further enquiries be conducted with the greatest discretion. It went on : 'We would warn you confidentially that Justus Perth is in fact Count Otto von Z . . ., a cousin of His Majesty the Emperor . . .'

Attached to the letter was a more recent photograph than that which I had received four years earlier and it was this which caused me to exclaim. Count Otto von Z . . ., the pseudo Justus Perth, was none other than Dhürmer, the clerk who had tried to make the two legionaries stake their claim !

My investigation was immediately simplified. Shortly after the original incident Dhürmer had been sent with the first draft to Tonkin. All I had to do was to telegraph Hanoi.

The reply came two days later : 'Legionary Dhürmer repatriated for health reasons. *En route* to Singapore.'

Another cipher telegram to the French consul in Singapore.

Reply: 'Legionary Dhürmer, alias Count Otto von Z . . ., has jumped his ship in Singapore and cannot be found.'

'I never unravelled the mystery,' General Bruneau concluded.

But it was a mystery only on the surface. By enlisting in the Foreign Legion Otto von Z . . . had turned his back on his past and for him there was no question of reversing his decision.

What other explanation can there be for leaving so great a fortune to unknown heirs?

So from Sergeant-Major Martin to pseudo Justus Perth the men of the Legion have made up their minds. For those who dream of a different life the Legion is a haven offering a unique opportunity to shed the 'old man' and acquire a new personality under another name.

The Legion is a haven. But it is more than that; to say that the Legion offers the drop-outs or the fugitives, whoever they may be, an opportunity to refashion their lives is an over-simplified explanation; not only is it inaccurate but it is incomplete. The Legion also forms part of these men's dream. The glory of the Legion is an essential part of their life. These unknowns, be they piteous or picturesque, heroes or faceless beings, have become 'sons of France not by the blood that they inherit but by the blood that they shed'.

Without these men a history of the Foreign Legion would be a mere survey, a list of dates, places, names and figures; it would be an inventory, taking no account of hardship, blood and death, nor of sand, mud, hunger, thirst, sweat or cold. It must never be forgotten that these men are the essence of the story. Though they may have put their past behind them, neither the military machine nor the white *képi* of the Foreign Legion has turned them into standardised automatons.

Today, as yesterday, the marching song of the 3rd Foreign

Legion Regiment is *Mein Regiment, mein Heimatland* – 'My Regiment is my Country'.

On 14 July 1946, shortly before his death, Georges Bohnert, an old ex-legionary of German extraction who had retired to Mauchenheim in the Palatinate after serving in the Legion for twenty years, drew up his will.

> Before my death I have still a few words to say to my former government.
>
> I wish to be buried shrouded in a tricolour flag and be escorted to the cemetery by a French military detachment.
>
> I bid a cordial farewell to the commandant, officers and men of the Foreign Legion.

Part One

Eugénie

1831–1900

Eugénie

Màrching Song of the 1st Foreign Legion

I

Eugénie les larmes aux yeux
Nous venons te dire adieu.
Nous partons de bon matin
Par un çiel des plus sereins.
Nous partons pour le Mexique
Nous partons la voile au vent.
Adieu donc belle Eugénie
Nous reviendrons dans un an.

II

Ca n'est pas facile du tout
Que de penser à l'amour
Surtout quand il fait grand vent
Par-dessus l'gaillard d'avant.
Nous partons pour le Mexique
Nous partons la voile au vent.
Adieu donc belle Eugénie
Nous reviendrons dans un an.

I

Eugénie, with tears in our eyes,
We come to bid you farewell.
We are off early in the morning
Under a cloudless sky.
We are off to Mexico;
We leave under full sail.
Farewell then, lovely Eugénie;
We'll be back in a year's time.

II

It is not at all easy
To think about love,
Specially when the wind blows hard
Over the fo'c'sle.
We are off to Mexico;
We leave under full sail.
Farewell then, lovely Eugénie;
We'll be back in a year's time.

Sidi Chabel

11 November 1832

'I'M THE sheik here!'

Captain Giudicelli was foaming at the mouth. He was an ill-tempered little Corsican and his rages were famous throughout the Legion, from Morocco to the Ardennes. He had never been able to address a legionary without a sprinkling of four-letter words. At this moment, perched on the edge of a shell-crater a few yards from the German lines in front of Verdun, he was shouting off his mouth, his high-pitched voice carrying like a cavalry trumpet.

Down below, at the bottom of the shell-hole, Kopeck, the legionary on duty, was hunched down, the collar of his sky-blue greatcoat level with his handlebar moustache. He was more afraid of his captain's rages than of the enemy's medium guns.

'Who the hell's landed me with a legionary like this!'

Much shouting had left Captain Giudicelli's mouth always open like a capital O; he would soon have been raging around for fifteen years. He was always in this state. He bellowed even when fighting and he fought well. No doubt this was why his legionaries adored him and would go anywhere with him. And they had gone too – since 1914 (three years ago), No. 2 Company of the 1st Foreign Legion (nicknamed the 1st House-movers) had fought everywhere, from Champagne to the Ardennes and the Marne.

'Wait till we're out of the line,' Captain Giudicelli yelled once more. 'I'll teach you your job, and with a boot up the backside too. I'm the sheik here.'

He stopped, not because he had calmed down but in order to draw breath to go on. Then, from the enemy trenches a few yards away, came a sarcastic voice: 'Well, if it isn't that bastard Giudicelli still shouting off his mouth!'

For a moment the captain was speechless, dumbfounded that opposite him there was definitely one of his ex-soldiers who had enlisted in his own country's army after serving his five years in the Legion.

Perhaps Captain Giudicelli was tempted to ask the man's name or he might even have done so instinctively; he could then have 'cornered' him even long afterwards, for Giudicelli had a memory like an elephant's. No doubt he realised that that would be a pointless question, but he could not resist having the last word. Meanwhile Kopeck peered up from the bottom of his shell-hole, anticipating the moment when someone would find a gap in the 'old man's' defences and put a sock in him.

'Isn't your captain at the end of his tether with a "legionary" like you?' Giudicelli shouted in a stentorian voice.

Then, more quietly and with an ill-humoured glance at Kopeck who had resumed his attitude of humility: 'One can never tell what these bloody foreigners are going to do next!'

France has always had foreigners in her service. Even at the time of Charles VII and Louis XI there was a Scottish Guard, immortalised in the novel *Quentin Durward.* Later, Englishmen, Irishmen, Germans, Poles and Swiss fought for France in their hundreds.

They were recruited as campaigns demanded and enrolled as occasion required. Initially they were professional mercenaries, serving the King's cause to the death. They formed independent companies under a captain or sometimes a general, himself of foreign origin – an example was Maurice of Saxony, one of the most famous eighteenth-century marshals. Other units, directly under the Royal House, were formed into regiments. On 13 July 1789 when the Swiss Guard defending the Tuileries was massacred by the good people of Paris, the Queen's Hussars, who were as Austrian as Marie-Antoinette and constituted the Esterhazy Hussar Regiment, stayed peaceably in their billets on the Place Vendôme – the business was no concern of theirs. The Saxony Dragoons, in barracks at the Palais Royal, did the same. Today we are entitled to wonder what would have happened to the infant revolution – and the Bastille – if they had chosen to take action. But they served a master, not a country, and they did not become involved in its internal quarrels. Few of them, however, went abroad to join the armies of the princes. They stayed where they were and three years later formed the cadre for the new Independent Foreign Legion set up by the Republic.

Events moved fast and the world was changing. On 7 June 1792, taking advantage of the deposition of Louis XVI, the Prussian army captured Longwy and then marched on Verdun. The country was in danger and, with its back to the wall, the National Assembly voted a decree on 1 August:

> Considering that circumstances require an increase in the armed forces, it is decided that:
>
> Under the authority of the Executive there shall be formed as soon as possible a new Legion to be entitled 'Independent Foreign Legion' and in which only foreigners can be enrolled.

Clearly events were in charge and France was at bay. The purpose of this law was to group under a single designation the various scattered regiments consisting of Germans, Irishmen or men of Liège. Rapid action was taken and on 20 September infantry and cavalry of the Foreign Legion took part in the battle of Valmy. It seems, however, that the foreigners were not over-enthusiastic, for on 3 August a law was promulgated granting considerable benefits – including an annuity of one hundred pounds – to foreign NCOs and soldiers who 'come to serve under our republican flag'.

The French Revolution considered itself more than a national movement; it was an ideology. It was not merely going to war; it was embarking on a crusade. So there were organised a Batavian Legion, a Germanic Legion, an Italian Legion, a Polish Legion. The future Marshal Augereau began his career as a lieutenant in the Legion.

It was an astounding phenomenon, this Legion, inspired with a revolutionary ideal – intended to permeate all Europe – and mobilised in the service of republicanism. Here was germinating the concept of an international army serving the cause of a political idea. Inevitably one compares it with that other army, formed much later for another crusade – the international units of the German Waffen-SS in the war of 1939–45; they were mobilised against Bolshevism. The terms used are different but the general idea is similar.

> If the kings close their ranks and despatch great armies against the free peoples, the latter must accept into their ranks all those men who feel themselves called by some sublime impulse to fight for the sacred cause of Liberty.

Europe was mobilised against barbarian tyrants and their 'white hordes', long before the 'red hordes'.

The Revolution ended; the Empire came into being. It inherited three Legions, relics of the Directory's decree of 22 Fructidor, Year VII. They were the Italian Legion, the Legion known as that of the 'Northern Franks' (they had not yet become Belgians) and the Maltese Legion which took part in the Egyptian expedition.

Under the Empire further units were formed: a Swiss half-brigade to replace the 'Swiss Guards', a Hanoverian Legion, a Portuguese Legion, a Spanish Legion and the 'Legion of the

Vistula'. They fought more or less everywhere, from Wagram to Eylau and including Spain. In 1815, renounced by the Restoration, they joined the ranks of soldiers on half-pay.

Good soldiers, however, seldom remain unemployed. Though scattered by the blast of defeat, the survivors of the European Legions reassembled – for a year only – in the 'Foreign Colonial' service and then in three regiments formed to 'occupy' the unemployed. Though disbanded on 6 September 1815, these regiments of foreigners reappeared, considerably reduced, under the title of the 'Royal Legion'. In 1821 (still further reduced – no war, no soldiers) it became the Hohenlohe Regiment.

The Restoration ended; the July Monarchy came into being. Meanwhile the affair of the Sultan's fan had led the French army to cross the Mediterranean. So while King Charles X was losing his throne, Algiers fell to General de Bourmont.

The Hohenlohe Regiment did not take part in the capture of Algiers. It was lying idle pending disbandment. It had been in existence for ten years and so far the life of the foreign regiments had been a short one. On 15 January 1831 a royal decree put an end to it.

Three months went by. On 9 March appeared a law – followed next day, 10 March, by a royal ordinance – organising a new Legion and laying down its charter under which all enlistments have been made ever since.

Paris, 10 March 1831.
Louis-Philippe, King of the French,
To whom it may concern, greeting.
Considering the law of 9 March 1831,
Considering the report of Our Minister,
Secretary of State of the Department of War
We have commanded and now command
as follows:

There shall be formed a Legion consisting of foreigners.
This Legion shall assume the title of 'Foreign Legion'.

It may seem surprising that France should owe to Louis-Philippe, the 'bourgeois' king, scion of a despised family and remembered by history only for his pear-shaped countenance, this élite unit which was so soon to create for itself a special place in the record of deeds of courage.

Courage did not at the time figure high on the list of civic virtues. The July Monarchy set itself up as pacifist and on Sundays the regimental bands of the day would grind out its favourite anthem: 'Where can a man be happier than in the bosom of his family?'

The English, though naturally distrustful, began to breathe again; France seemed to have lost her belligerent inclinations. Having pondered the matter for a hundred years, they now suspect that it was Marshal Soult, Wellington's opponent in the Peninsular War, who gave Louis-Philippe the idea. It is possible, for Soult was generally considered a spiteful old man. However this may be, having taken a pace forwards, the next step was retrograde: the Foreign Legion could not be used on the continental territory of the Kingdom of France.

This backward step was to govern the Legion's programme. In France it could only parade. If sent out of France, however, so much the better – that would be in order to fight.

The conquest of Algeria had been under way for a year. After initial victories the French expeditionary force had been compelled to restrict its operations as a result of British pressure and Louis-Philippe, who set much store by his reputation as a convinced pacifist, had withdrawn the majority of the troops.

Lamoricière, the commander-in-chief, was reduced to improvisation. Nevertheless he persisted, concealing his advance under innocent-sounding phrases such as 'pacification' and 'field firing exercises'; he found ways and means of evasion, establishing command posts and pacification centres known as 'Arab Offices'.

There were two opposing schools of thought. On the one hand the soldiers considered it impossible to 'hold' Algeria unless the hinterland was under control; the central authorities, on the other hand, were unwilling to take risks outside the coastal towns and ports. So everyone was playing a double game. Even as late as 1903 Lyautey was still doing so; he even went so far as to rechristen towns in order to explain the presence of his troops. The following instance is famous: the Minister ordered: 'You are forbidden to move to Bechar.' 'Me?' Lyautey replied. 'I have no intention of going to Bechar. I am at Colomb.' Naturally he did not explain that he had himself given the name of Colomb to the post at Bechar.

During this period orders from Paris took days to arrive. The military men accordingly acquired the habit of acting on their own initiative and were quite prepared to disregard the rules of so-called discipline. From this period grew the Foreign Legion's customary attitude of independence which ultimately became one of its most cherished traditions – a superb and resilient attitude of nonchalance towards the hierarchy. Colonel Maire, one of the greatest figures of the Foreign Legion, used to say as he rose from his chair: 'Those who don't like it can still find room between my seat and my arse.'

But this was said in 1925. We return to August 1831, when the Foreign Legion disembarked in Algiers. There being nothing to defend, they had come to conquer.

To start with their uniform. For the moment it was the regulation uniform worn by the French army everywhere: crimson trousers, royal-blue tunic buttoning across the chest, iron-grey cloth cape – this was the month of August, be it remembered! – *képi* with crimson band, blue sash and piping, pipe-clayed buffalo-hide accoutrements. The only feature distinguishing the Legion from normal infantry was their buttons. In addition to the five-pointed star they carried the words, invariably engraved in capital letters, 'LÉGION ENTRANGÈRE'.

The new unit consisted of seven battalions, each by tradition formed of soldiers of the same nationality. In this case three battalions were German, one Spanish, one Italian, one Belgian and one Polish. These seven battalions were distributed among the three coastal provinces, five to Algiers (Nos. 1, 2, 3, 5, 7), one to Bône (No. 6) and one to Oran (No. 4). For the next four years the decision to keep nationalities together served the authorities well; command was easier since officers and NCOs spoke the same language. The disadvantages only appeared later when the legionaries took sides in the continental quarrels of their far-off home countries and so introduced their national squabbles into the Legion. Not until five years later, with the arrival of Colonel Bernelle, was homogeneity enforced. The measure proved its worth in the end since, to make themselves understood, the legionaries were forced to use French for everyday communication. Alternatively, as they still do, they used a sort of European patois, an esperanto ahead of its time, in which the most hard-worked word was the one dear to the French army – '*démerdez-vous*' ['Get on with it].[1]

A notice in the NCOs' mess reads: 'As a matter of courtesy people are asked to speak French in the mess.'

So the Foreign Legion 'got on with it' in Algeria. It had to start from scratch and its first action was to adapt its uniform to the climate. The first innovation was the introduction of white overalls. Initially this was a plain outfit in raw hemp, rough and greyish brown in colour, similar to packing material or sackcloth. The legionaries took a certain pride in making it as white and as soft as a choirboy's surplice; one of the old hands says: 'By using elbow grease and chlorine filched from the quartermaster, one could produce trousers bearing no relation to sackcloth and looking, in the sunshine, like an imperial general's breeches.'

The Imperial period was much in the mind of the Legion.

1. '*Demerdieren*' in German, '*demerdar*' in Spanish, '*demerdovar*' in Czech and so on.

Many of the officers had served under Napoleon and, in bivouac in the evening, would tell stories of the Grand Armée. So did the 'old hands'.

Among these old hands was Colonel Stoffel, the Legion's commander, and Mollembeck who succeeded him. Among the youngsters were several future marshals: Saint-Arnaud (a captain at the time of the capture of Constantine), Bazaine, later held responsible for the French capitulation at Metz during the Franco-Prussian War, Espinasse who was killed at Magenta as a general and Villebois-Mareuil who became the Legion's historian and died in South Africa fighting for the Boers.

The Algeria of 1831 bore little relation to what it later became. It was a savage desert country, unhealthy and smelly. Its wild vegetation lent itself to the ambushing of troops moving about the country once they had left their fortified posts or the plains.

In addition to the normal duties associated with active service – organising bivouacs, staging posts and supply depots – the Legion had to guard long columns; it was too strong to be attacked but too itinerant to pacify the country. There were the same problems everywhere: villages to be captured and then abandoned next day, tracks to be opened up, villages to be built.

Bugeaud's idea was that territory of no military interest should be held by colonists. The Foreign Legion protected them as they established themselves and helped them to build their houses. The colonists, who came mostly from Sicily or Spain, wore black boots which earned them from the Arabs the pithy nickname of '*pieds-noirs*' [Black-Feet], sometimes used by the colonists as a title of distinction.

But there were other duties to be undertaken in this country. Pacification could not take place without war, and warlike operations became more numerous. This was inevitable; as they worked their way slowly into the country the colonists and soldiers encroached on native territory. The natives were indignant and attacked the weaker detachments, although, for good measure, they then attended the funerals of the victims. These attacks led to repression and repression led to revolt. For a time revolts were localised. Large-scale revolt began only when Islam produced a nationalist prophet, the Emir Abd el Kader. Soon all the animosities which had built up since the days of the Turks became focused on the French. Abd el Kader was all the more likely to be listened to because he proclaimed a Jehad – a holy war.

Algeria was alight from Oran to Bône.

The Foreign Legion set the tone. Since the Arabs were on the move, the generals thought, why not take advantage of the fact to push French conquests further?

On 27 April 1832 the 1st and 3rd Battalions of the Foreign Legion captured the redoubts defending Maison Carrée, only a few miles from Algiers. The operational boundary had been pushed a little further east. The same thing happened in the west. The 4th Battalion – Spaniards – captured the port of Arzew from the Emir's followers and then took the post of Mostaganem in their stride.

These were the Legion's first feats of arms. It was rewarded with a flag – its first colours, presented on 24 June by Combes, the new colonel. The inscription read: 'The King of the French to the Foreign Legion'. Such exploits as these – and the publicity given to them in Paris by the generals commanding in Africa – had been required to remind the man who called himself 'King of the French' that foreigners were fighting for France – and without a flag.

Abd el Kader's rebellion spread. On 11 November 1832, however, after a series of local victories, he finally found himself facing an opponent worthy of him – the Foreign Legion. It was waiting for him at Sidi Chabel, at the gates of Oran.

Abd el Kader arrived before Oran at the head of 3000 cavalry and 1000 infantry. His position was impressive; he held the whole line of hills overlooking the city, in particular the Djebel Tafaraouini where was the shrine of Sidi Chabel.

The French under the command of General Boyer decided to attack. It was a widespread frontal offensive. An initial charge by the Arab cavalry was halted by the French artillery and then the Chasseurs à Cheval and the Foreign Legion counter-attacked. On the left of the line the 4th Foreign Legion Battalion was thrown forward. It split the enemy front and caught Abd el Kader between two fires.

The legionaries proved tough under fire. They were almost all Spanish ex-guerrillas, survivors of the campaigns of the Imperial period, and sabre charges did not scare them. They knew how to unseat a cavalryman, how to break into the middle of the squadrons and skewer either the horse or the man. The climate was nothing new to them; they were both cunning and tough. The Arabs were impressed and were put to flight by the Legion's dash and flexibility in manoeuvre.

That evening Abd el Kader withdrew. Oran had been saved.

Early in the rebellion a second battalion of the Legion was detached to join No. 4 at Oran; this was the 5th, consisting of Italians.

For nearly two and a half years the life of the Legion alternated between battle and truce, pacification and construction. There were few major engagements, merely one or two ambushes, but few results. The tactics prescribed by Bugeaud were simple, similar to those of the Roman legions – *ense et aratro* – 'by the sword and the plough'. The plough furrowed deep, the sword struck hard. There was no parleying with the rebels and Montagnac, commanding the Chasseurs à pied, who was killed in 1845 at the famous battle of Sidi Brahim, went so far as to write (to the revulsion, be it noted, of the majority of officers in Africa): 'One severed head produces more terror than the death of fifty individuals.'

The war entered a guerrilla phase. Abd el Kader regrouped

his forces away from the Atlas Mountains and awaited his moment.

On its side the Legion also waited – for departure.

In January 1835 the Legion was withdrawn from operations to prepare for a fresh campaign elsewhere; precisely where was not known for the moment.

A tripartite agreement existed between France, Portugal and Britain. As a result King Louis-Philippe was led to promise substantial assistance to Marie-Christine, regent of Spain, in order to keep her and her dynasty, represented by the young Isabella, on the Spanish throne. The Infante Don Carlos, the pretender, was supported by the Basques and Catalans who had risen in armed revolt. French assistance took the form of hiring out soldiers, and naturally the troops so hired were the Foreign Legion battalions.

Louis-Philippe's government was undoubtedly liberal by the standards of the time but its broadmindedness did not extend to today's ideas about equality and the rights of the individual. The two dukes, for instance, de Broglie for France and de Frias for Spain, who together settled the arrangements for the move of the Legion, seem to have looked upon it as if it were on a par with those mercenaries who sold their services three times a week to the Pope, the princes or the Republics in fifteenth-century Italy. Even accepting the fact that the Spaniard had no reason to be solicitous for the Legion, the French representative ought to have insisted on a proper agreement.

Such as it was it ran: 'On the proposal of His Majesty the King of the French, Her Majesty the Queen Regent has resolved to take into the service of Spain a body of foreign troops at present in the service of France.' Reading between the lines, Louis-Philippe was simply using the Legion as a temporarily recruited body of men to whom his only obligation was to pay them. There was no question of loyalty.

Article 3 of the convention signed in Paris on 28 June 1834 read: 'The King of the French undertakes not to recall this force to the service of France unless Her Majesty the Queen Regent has given formal consent.' This was undoubtedly a disturbing paragraph. The problem eventually solved itself as the contracts signed by the legionaries themselves expired.

The terms of this 'grant of assistance' soon became known and, as may be imagined, the officers of the Legion were not best pleased.

A great gulf was fixed between Paris and Algiers. Already the battles and hardships of the Algerian campaign had given rise to that which was subsequently to become the very foundation of the 'spirit of the Legion': unlike the mercenaries of other days who merely served a master, the legionaries had begun to consider France, and through her their regiment, as the country of their choice.

'There is clearly no question of a refusal to volunteer for Spain,' the commander-in-chief in Algeria, a certain General Rapatel, said to the Legion's commanding officer.

'Very good, sir,' replied Bernelle who had just arrived to command the Legion.

Bernelle was an astounding character. He was an impressive figure with a taste for panache inherited from the Empire; he had taken part in all the Napoleonic campaigns. For him the move to Spain meant that he would really be the 'boss' of his unit, the only Frenchman on foreign soil. Then he was flattered by a proposition put to him by the minister: he was to be appointed 'camp marshal' in the Spanish service – a sort of five-star general. Admittedly this looked rather like a temporary rank only, but what senior officer would not be tempted by so sonorous a title even if it was only provisional?

So Bernelle did his utmost to convince his officers. They were bitter and said so. The legionaries were bitter too, but they held their tongues; according to a remark which later caught on they had 'signed on in order to sweat blood'.

'The war in Spain,' so the story went, 'is only a campaign like the rest; acts of bravery will be rewarded; you will win decorations, medals and . . . officers' stripes.'

This effort of seduction was nullified by a blunder on the part of Marshal Maison, the Minister of War; with conspicuous lack of common sense and a narrow view of the meaning of the 'loan' of the Legion, he coolly ordered that: 'The Legion will hand in its flag and the officers will cease to wear the tricolour cockade.' In his eyes it was all quite simple and the matter was settled. France was now rid of this foreign unit which seemed to him to have no place in the army; he did not understand its usefulness, though he feared its spirit.

There was an immediate come-back. Major Kléber preferred to leave the army rather than cease to wear the French colours. Six captains and eleven lieutenants followed him into retirement. The remainder resigned themselves willy-nilly to sharing the lot of their Colonel Bernelle, undoubtedly from a sense of discipline rather than conviction.

Other orders were purely administrative; new packs were drawn, arms and equipment brought up to full strength. In short, all preparations were made for movement overseas. This was to take place in two stages, troops in Algiers province leaving from Algiers and the rest from Oran. The fleet would then concentrate at Palma, Majorca, and the whole formation was then to disembark at Tarragona.

The 1st, 2nd, 3rd and 6th Battalions paraded in Algiers before General Count d'Erlon, the Governor-General of Algeria. They embarked on 27 June. The squadron sailed on the 30th.

The two battalions from Oran were behind schedule. They had been recalled to operations as a result of an unexpected offensive by Abd el Kader. On 26 June he had attacked French columns under the command of General Trezel in the Moulay-Ismail defile.

No better choice could have been made for an ambush. For twenty hours the French were subjected to repeated assaults by the Arab light cavalry. The position became untenable and fighting soon became desperate.

On the evening of 27 June General Trezel sounded the retreat. Only one route was open to him for his withdrawal –

through the Macta salt marshes between Arzew and Mostaganem.

At this point Abd el Kader was hoping for total victory over the French. All day he had been manoeuvring to block all the other exits. By the evening of the 27th he knew that he had won.

What he did not know was that Trezel now had available two battalions of the Foreign Legion which had been alerted as a last-ditch measure and had reached the French lines by a forced march, moving towards the sound of the guns. These two battalions, the Italian and the Polish, went into action at once. To cover the withdrawal they made an initial attack on the evening of the 27th. It was carried out with spirit and dash.

After a single burst of musketry they formed square and advanced on the Arab cavalry. Then, lowering their bayonets, they drove in shouting. For a moment the enemy relaxed his stranglehold. Trezel was able to get away, although a few miles further on he became bogged down in the Macta marshes.

That night the Foreign Legion formed the rearguard of the column, taking its dead with it. Two officers had been killed leading the battalions – Lieutenant Josefovitch of the 4th Polish and Lieutenant Boldini of the 5th Italian.

Trezel's withdrawal, however, gained him only a short respite. Struggling through the marshes, his infantry moved slowly, still encumbered by supply wagons and the wounded.

On the morning of the 28th Abd el Kader, thinking the annihilation of the French to be within his grasp, launched his cavalry against the long column which was snaking its way through the reeds. His tactics were to split his enemy into several sections and then reduce them piecemeal – classic guerrilla tactics which the Vietminh were to rediscover a hundred years later in Indo-China.

Major Horain, commanding the 4th Foreign Legion Battalion, was quick to see the threat; he was still acting as rearguard and, moving up the column, he formed his men up for the attack, acting as a screen between the exhausted infantry and the enemy front.

'*Vive la France*,' the legionaries shouted with one voice.

Horain was a brave man who feared neither death nor bullets. Hardly had he started to move than he fell, his horse having been hit by five bullets. But this did not deter him; he continued the attack on foot, sabre flashing. His legionaries gathered round him, led by 2nd-Lieutenant Bazaine.

By midday the battle was raging in blistering heat, to which was added the fumes rising from the stinking water, almost boiling in the pitiless sun. Already the Arabs had succeeded in penetrating the French line at its weakest point. They reached the ambulances, overturned them and massacred the wounded, frequently running them through with their own bayonets. They burnt the forage carts and supply wagons. Trezel at the head of the column, knowing nothing of this massacre, urged his men on. The column was split and those who lost contact were doomed to destruction. On the flanks the Legion was still fighting.

The 5th Battalion (the Italian) was the first to receive news of the enemy attack on the supply train. It disengaged, joined the rearguard and charged repeatedly, forcing the Arabs to fall back. Then night fell. Guns at the ready, the battalion stood its ground, giving the convoy time to withdraw and join the head of the column. Only then did the 5th Foreign Legion begin its own withdrawal; it went on all night, fighting continuously. It was supported by the 4th Battalion, the two units leapfrogging.

By the time Trezel reached Oran the sacrifice made by these two battalions of the Legion had already become famous, almost legendary. As they sailed towards Spain and Tarragona the legionaries evoked astonishment and admiration in Paris.

'What?' certain generals said to themselves, primarily those in headquarters who are invariably sceptics; 'are these drunkards, thieves and bandits really exceptional soldiers as well?'

The surprising fact is that, whatever the period and whoever the general, there is always someone in Paris who has to make the discovery that the Foreign Legion fights well.[1]

Meanwhile the French squadron concentrated at Palma on 10 August. Six days later the Foreign Legion landed in Spain.

1. M. Vincent Monteil, author of a book entitled *Les Officiers* and otherwise a brilliant strategist, has accomplished the feat of telling the story of a hundred years of French warfare without once mentioning the Foreign Legion!

Terapegui

26 April 1836

'WHO GOES there?' shouted the symbolic sentry at the gate of Tarragona, and Bernelle, erect and magnificent on his horse, replied in stentorian tones: 'Spain! Foreign Legion!'

Colonel Conrad, deputy to the commander and a Prussian officer, whispered to Lieutenant Hebich: 'Lucky he said "Foreign Legion"; we might have been taken for Spaniards.'

Hebich smiled. Like most Germans in the Legion he had a weakness for Conrad, who was nicknamed 'old Fritz'. In fact, however, Conrad was an Alsatian. He too was a veteran of the Napoleonic campaigns. He had been wounded several times and had been in the Russian campaign, from which he had returned not only with a number of additional wounds but also with a wife, daughter of a counsellor at the Tsar's court. He had also served under Masséna during part of the Peninsular War and the country held no pleasant memories for him. Better than all the rest he knew that war in Spain was seldom merry and bright.

Behind him rode von Rosen, the quartermaster-sergeant. He was another German, a genuine count, a noted drunkard and the campaign's historian. In addition to one or two well-executed portraits, the Legion acquired from him its reputation for heavy drinking.

'Conrad,' von Rosen records, 'was a small broad-shouldered man of about fifty with a good-looking, unmistakably German face. Both his attitude and his attire were truly soldierly. He had a full-length beard, giving him a martial air; on his head he wore a red cloth gold-embroidered cap without a peak.

He rode a snow-white arab, so could always be picked out from a distance in the midst of the fighting. For this reason the Spaniards nicknamed him "The brave one on a white horse".'

This is clearly an appreciative description despite the somewhat jingoistic overtones.

Once the headquarters group had passed, Tarragona welcomed the Legion with cries of 'Up the foreigners! Up Algeria'. There were also one or two shouts of 'Up with liberty' from republicans, supporters of Riego who had been executed some years earlier. Some people sang the Marseillaise, which no doubt pleased the officers and equally certainly displeased Spanish officials, civilians and clerics. The loudest cheers were evoked by the bearded sappers with their shouldered axes, whom Bernelle had turned into his bodyguard. As already mentioned, he had a taste for panache, and was to gain a reputation for putting on a show.

Bernelle was a great admirer of Napoleon and tried to imitate him in everything, including his sense of family solidarity. He found places for six of his relatives in the ranks of the Legion, including his brother, a diseased invalid who did nothing other than draw his pay and sometimes warm his brother's heart when the latter was particularly scandalised by the behaviour of Madame Bernelle.

Her series of nicknames speaks volumes – Isabella III (Isabella II being the Infanta for whom the Legion was fighting), the Princess of Navarre or – less politely but more accurately – 'The Whore'. She had her husband completely under her thumb. Though she would not deign to consort with the great bearded legionaries of her husband's bodyguard, she liked to be surrounded by 'good-looking, pleasing young officers', so von Rosen says. He also records some of the rhymes made up about her by the Legion, one of the more printable of which runs:

Madame Bernelle avait ordonné
Que les plus beaux garçons soient galonnés.

Madame Bernelle had commanded
That the best-looking boys be commissioned.

In fact, though she may have handed out officers' stripes, she was also generous in dishing out days under detention to the legionaries, perhaps because they had watched her sunning herself in her garden in her négligée or because they had not shown sufficient respect to her.

Von Rosen, who has a fund of stories about her, records that one day a drunken legionary was crossing the road just as 'La Maréchale's' carriage passed. The horses bolted, the carriage overturned, and there was Madame Bernelle in the ditch, naked to the waist. 'Eight days' detention,' she yelled even before readjusting her attire.

Colonel Bernelle was not apparently hampered by his wife's indiscretions. He put all his energy into the organisation of his unit, with a little brutality and perhaps some over-enthusiasm, but he welded it together to an extent never before known. His first measure designed to achieve solidarity was to change the composition of battalions. Next he instituted an iron discipline to which he personally contributed by innumerable blows from the stick which he invariably carried. His ways produced some grumbling but they reinforced the legend of the strong, just, commander. This became the standard picture of a Foreign Legion commanding officer – undisputed master of his unit and totally unmanageable in all other respects. This was the way the Legion liked them.

So, on 22 August 1836, a new Legion took the field. It comprised 123 officers, 4021 NCOs and men.

Two columns were formed, one of three battalions under the orders of Conrad and another in reserve direct under Bernelle. Their orders were to attack and bring to a halt the Carlist army which had emerged at a strength of 5000 men from its strongholds in Upper Aragon. On 23 October Conrad reached Lerida, where an unhappy surprise awaited him. The Spanish headquarters to which he was responsible split up his force and distributed his companies as reinforcements to the Spanish units – the '*christinos*' – facing the advancing enemy. In fact, and this was to be the Legion's fate in Spain, for most of the time these scattered companies carried the whole weight of the fighting on their own. Conrad himself had bitter experience of this. He was later killed because, having shouted 'Advance', he found himself alone with a handful of legionaries; the Spanish Guards had not followed him.

Nevertheless, from the outset of the campaign, the Legion distinguished itself. The enemy was not yet familiar with it and pushed forward boldly.

On the morning of 16 September a thousand Carlists tried to force their way through a defile held by a small force of the Legion under Captain Ferrary. He was subjected to an attack lasting sixteen hours but did not give way an inch. And then in the evening, amazingly, he gave the order: 'Advance! Charge!'

His three companies – three hundred men – charged, and cleared the area in a few minutes. But this victory was not enough for Ferrary. He pursued the enemy, hunting him along, and forced him to take refuge in Guimera Castle. Ferrary invested the place. Four days were enough for the Carlists to realise that they were trapped. They surrendered. Then, in cold blood, Ferrary had eighty guerrillas shot – peasants who had joined the Carlists. They were not in uniform.

Wars in Spain are seldom chivalrous affairs and the Legion was one of the first to learn the lesson. A few days earlier a small post manned by thirty legionaries under 2nd-Lieutenant Dumoustier was attacked by three hundred of the enemy. After several days' siege the garrison had fired off all its ammunition, exhausted its supply of water and so sallied out with the bayonet. It broke through but was then betrayed by its guide. The enemy caught up with it and surrounded it, this time in open country. Being without food or ammunition Dumoustier was forced to surrender.

'Join us,' the Carlists proposed. 'We have room for men

like you in our ranks.' Naturally this was met with a refusal.

Then, for three days, Dumoustier and his men were paraded naked through the rebel villages, with ropes round their necks, and were treated like cattle. Finally, not knowing what else to do, the Carlists shot Dumoustier and three of his NCOs. The others died a miserable death. Only one or two legionaries managed to reach the French lines and tell the horrible story.

During its four years in Spain the Legion was connected, in one way or another, with a long series of atrocities. Don Carlos, the Pretender to the throne, had hoped for a time that he could assert himself by force of arms. The arrival of the Legion constituted an unexpected reinforcement to his enemies, making the victory which he had anticipated questionable. So, although he had accorded the privileges of the 'laws of war' to the majority, from the outset he excluded the Foreign Legion.

If he thought that he could demoralise the Legion, he was mistaken; it met savagery with fortitude and treachery with courage. Its discipline counter-balanced numbers. It must be admitted, however, that it was adept at applying the biblical precept, 'an eye for an eye and a tooth for a tooth'.

On 24 April 1836 'Colonel-Marshal' Bernelle was warned by his spies that a large Carlist force was on the move in the area of Zubiri, only a few miles from his position. The 4th Battalion, commanded by Lieutenant-Colonel Magnier, was ordered to bar the road to the enemy column. The order was to 'intercept and destroy'.

The weather was terrible. A deluge of hail, soft snow and icy rain was descending on the foothills of the Pyrenees – the Legion was in Navarre and, though the calendar may have said that it was spring, this news had not penetrated to the high Spanish plateaux. Battling their way through the squalls, the scouts of the leading company, commanded by Captain Aneth, worked their way up the hill and reached the plateau. Facing them was a small farm named Fernandorena, a dark lonely building in keeping with the sinister countryside.

Aneth suspected the place and ordered: 'Deploy into line.' He was right. Hardly had the first legionaries reached the outskirts of the farm than they were met with steady musketry fire. At the same time, out of the storm, a squadron of Carlist lancers emerged from a nearby wood, taking the legionaries in flank.

At once there was a real mêlée. The legionaries faced both ways, but the infantry installed in the neighbouring buildings were shooting almost at point-blank range, and losses were terrible.

The company's leading platoon was entirely encircled by cavalry. Its commander, Sergeant Samuel-Benoit Berset, who had served in one of the Swiss regiments under the Restoration, faced them like a lion. Several times he ordered a bayonet charge against the horses. But Aneth's legionaries were outnumbered and his ranks began to thin. Berset attacked again, his sabre carving a way through the lancers. At times they withdrew, only to return with greater fury. Berset hacked them in pieces. With bodies all around him he continued to fight.

Suddenly came a shout – a long-drawn cry heard above the noise of battle and the howling of the storm: 'Hurrah! Advance!' It was the rest of the battalion coming forward at the double and taking up position on the plateau. The Carlists realised that they were now no longer on top and retired forthwith. The legionaries found the body of Sergeant Berset and could hardly believe their eyes. He had been bayoneted fifteen times and thrust through three times with a lance; for good measure his body also bore two sabre-cuts, twenty wounds in all.

Mad with rage, his comrades swore to avenge him. They set off in pursuit of the enemy who scattered on the slopes of Zubiri; the legionaries' fury was such that they caught up with them. They killed 200 during the pursuit and brought back thirty prisoners who were shot on the spot. The Spanish answer was to burn alive five legionaries whom they had captured and managed to take with them.

Winter ended and spring arrived. According to orders from General Cordova, the commander of the Spanish royal army, the Legion was to hold the front 'from Pamplona to the Pyrenees'.

'Why not?' Bernelle asked, having been recalled to his duties as mercenary; he was to occupy 'fortified' posts. There were no fortified posts, nor was there any pay and, if one expects obedience, one should keep one's contracts. The legionaries protested. They were all the more dissatisfied in that, after a year of campaigning, no one had renewed their wardrobe and war is hard on uniforms. 'Keep your greatcoats,' they had been told. The great advantage of a greatcoat is that it hides the rags and economises in uniforms.

Above Pamplona Bernelle had located his main body in the town of Larrasoaña. All round it, facing the hills from which the enemy might emerge, the legionaries had established a line of blockhouses. The Carlists had not been defeated but they were confined to their own territory. Every day small groups of them arrived, hid in the woods and took pot shots. The French sentries could not reply since the enemy remained invisible. But they supposed that their time would come to take the offensive.

Bernelle thought so too. On 25 April he took the bull by the horns and sent a large working party into the forest to fell the trees. In a single day, working hard, the legionaries levelled the hill of Terapegui, from which most of the firing had come.

Next day about a hundred Carlists paraded in full view in front of the fallen trees and took up position facing the blockhouse. Taken by surprise, Bernelle sent out three companies to reconnoitre.

'Simply observe,' he told them, 'and under no circumstances start a battle.'

Gradually the hills became thick with enemy. They took up

position as if round an arena and, shouting insults, faced the legionaries who formed square in the plain. For two hours the legionaries remained impassive and did not reply. The Carlists became bolder; they came closer and, still shouting, surrounded Bernelle's two hundred men who stood there fearless, though with singing ears.

Then suddenly they had had enough. Without realising it the Carlists had tried the patience of the French too high. Levelling their bayonets the men moved forward, quickened their pace and then charged.

The Carlists were ten to one. Bernelle, who was following developments from his command post, had to do something. He assembled four hundred men, four guns and an artillery officer and sallied forth.

As if this was what they were waiting for 3500 yelling Carlists moved in to the attack. Their target, of course, was the 'camp marshal'. Several times they reached him and fired. Each time one of the bearded sappers moved forward and shielded Bernelle with his body. They numbered sixty and by the evening twenty-two of them had been killed.

Down below, on the edge of the Terapegui gully, the hand-to-hand fighting was frightful. Lieutenant Rousset, the artillery officer, was attacked by a particularly savage and ferocious gang commanded by La Sanca, and was forced to extricate his guns using his sabre.

The battle raged for six full hours. It was real butchery, continuous hand-to-hand fighting, 4000 men disembowelling each other on a narrow plateau and its adjoining slopes using cold steel, muskets and guns. Three times Captain Ferrary's companies sallied out of Larrasoaña and took the Carlists in rear. After firing their first volley the legionaries, furious at their losses, did not give themselves time to recharge their flintlocks but charged with the bayonet. All Bernelle's battalions were now engaged and he manoeuvred them primarily to enable them to make bayonet charges.

Finally Carlist morale cracked: they could not understand

why the legionaries clung to a soil which was not theirs. They withdrew, leaving over a thousand dead on the field. The 3rd and 4th Battalions of the Legion had lost 300 men.

The next day the Carlists returned bringing back the wounded prisoners. They had cut off their lips, pierced their eyes and torn off their eyelids. It is hardly necessary to add that from this moment there was was no question of ever taking a prisoner again in Spain.

In the light of their hardships Bernelle was determined to get his men at least some pay and some clothing, the promised promotions and decorations not having materialised. He became angry and wrote to Paris, but all he achieved was to fall out with the French Minister of War! So the great commander of the Legion retired disillusioned, and was relieved of his office. He had at least the satisfaction that he had not given way and he was fortunate in being spared the spectacle of the misery of his troops.

For there now began a prolonged agony. Nevertheless a commander was required, even though he was powerless to improve matters. 'Old Fritz' was still there. He was no pen-pusher nor diplomat, simply a soldier who reacted like a soldier.

'You've nothing to eat?' he asked his subordinates. 'Very well, we will go and help ourselves.'

He despatched expeditions into the villages, choosing Carlist ones for preference. Morale rose at once and the legionaries now felt better equipped to fight an enemy whose wives they knew intimately.

Winter returned, worse than the last. Nevertheless, though in rags and barefoot or shod with any old clout, the Legion still held its fortified line.

Sensing that their enemy was on the downgrade, the Carlists became bolder. They even suggested to the legionaries that they desert. Surprisingly and to their eternal credit, they remained staunch. Over a period of two years there were only eighty-four desertions.

On 10 March 1837, under the pressure of French public opinion – journalists had been describing the 'miseries of the Legion' – Marshal Maison, the Minister of War, despatched a consignment of clothing. After eighteen months of fighting this was an unheard-of luxury.

Morale rose by leaps and bounds. By the 15th the Legion was once more ready to fight. It was despatched to Larrainzar, a small village in a valley overlooked by a wooded ridge. Conrad, an experienced old soldier, considered that it was not possible to hold the village without occupying the heights. Disregarding his orders, therefore, he sent Hebich's company up to build a redoubt.

Hebich had the worst reputation of any officer in the Foreign Legion. He was a German, born in Ulm in 1793, and had fought for Napoleon in the Württemberg Guards. And he had not spared himself; he had been twice wounded at Polotsk in 1812 and received two sabre-cuts in the face at Juterbock in 1813. He fought in the French campaign. Then he deserted and went to Greece, where he was wounded again fighting for Alexander Ipsilanti. He deserted once more and returned to France.

Hebich joined the Foreign Legion in 1831. He was unpopular from the outset. He was an undisciplined rebel and came to blows with Lemaire, his captain, when he said that his men were a lot of 'good-for-nothings'. At Maison Carrée he announced that the Legion was 'nothing but a collection of sharks who would be prepared to make the understraps of its leggings out of the moustaches of crack units'. He was placed under arrest.

Colonel Combes, the Legion's first commander, said of him: 'This man is a disgrace to the rank of officer.'

'He is a boor,' Bernelle confirmed.

Conrad went even further – 'Unworthy of officer rank.'

Nevertheless Hebich fought well. He was hit by a bullet in

the head during the attack on Bougie and by another in the left leg at Sanguessa in the previous year.

Such was the officer who, on 26 March 1837, was to save the Legion.

The Carlists had decided to attack the French in Larrainzar village. They massed on the hills in complete ignorance of Hebich's position. In the morning Hebich could see how they were disposed. Without a moment's hesitation he opened a hellish fire against the enemy flank from his half-finished redoubt. The enemy were taken by surprise for a moment, but then realised that they must capture the redoubt if they were to have any chance of reaching the village. But they did not know what they were up against. Hebich gave way not an inch. He even made sorties to gain more elbow-room.

But he was suffering casualties. Lieutenant Sohacki, his second-in-command, fell dead with a bullet in the forehead. Hebich crouched behind the body using it as a breastwork. The legionaries copied him. Faced with such stubborn resistance, the Carlists redoubled their efforts, but at midday Hebich's redoubt was still holding out.

Alerted by the firing, Conrad sallied out from Larrainzar with the 2nd Battalion. He moved up the gully, took the Carlists in flank and relieved the redoubt. The 1st Battalion also 'marched to the sound of the guns'. It arrived an hour later, in time to take part in the final phase of the battle. The enemy decamped.

By his initiative, his courage and his stubbornness Hebich had prevented Conrad being caught in Larrainzar. By holding out, unsupported, for over three hours he had made the counter-attack possible. Undoubtedly this officer, whom no one liked, had saved the Legion.

But the Legion was on the way out. No reinforcements arrived and every month four hundred men departed on termination of engagement. By June 1837 Conrad was down to a single battalion, so he was withdrawn from his 'fortified line' and sent to operate alongside the Spanish army. In fact he went to die.

The Queen's army was stretched out in Aragon, an arid country criss-crossed by little valleys and ravines. It was in sight of Barbastro when several squadrons of Carlists attacked its right flank. The flank, formed of infantry and cavalry, was disorganised by the sudden shock. It scattered, and suddenly the Carlists had reached the centre of the army and found themselves faced by the 2nd Spanish Guards Regiment – and Conrad's Foreign Legion.

Surprise has always been the guerrilla's weapon. The Carlists made good use of it and pushed forward, as shock troops, their own Foreign Legion, a battalion of German and Swiss mercenaries.

The scene has become famous. The two legions faced each other, at first gossiping, recognising each other, exchanging news of old friends. Then they suddenly rushed upon one another, killing like madmen, using first their muskets, then bayonets, sabres, fists and fingernails. The Carlists came to the rescue while the Spanish Guards fled, leaving the Legion alone. Then, assembling the last remnants of his battalion, Conrad appeared, lifted his famous cap on the point of his sabre and shouted: 'Advance!'

At first the Carlists were driven back but a bullet hit Conrad in the middle of his forehead. He slid from his horse and collapsed into the arms of his ADC, Captain Bazaine. The commander was taken to the rear, Bazaine ordering: 'Say that the general is wounded' – he was afraid that the Legion's morale might collapse.

Even this subterfuge was useless, however. The battle was lost. At nightfall the legionaries were still firing to cover their men collecting the dead and wounded – a duty which the Legion invariably does its best to fulfil.

At midnight, in their bivouacs, they learnt of 'Old Fritz's' death. Mad with rage, they rushed towards the lines of the

2nd Spanish Guards Regiment, whose flight they held to be responsible for the old man's death. Only with great difficulty did their officers restrain them.

Reduced almost to nothing, the Legion occupied Pamplona for a time. Finally on 8 December 1838, finding that it still existed, the French authorities sent an order for its 'return to France'. So, in little groups, haversack on back and on foot, the survivors of the Legion left Spain. They numbered no more than 500 out of the 5000 men who had landed at Tarragona three and a half years before.

During their march 'home' they rubbed shoulders with a number of Carlists who, like them, had had enough of the war in Spain. Among them was a certain Martinez, said to be a former lover of Queen Isabella, who had gone over to the Carlists in despair. Though previously a formidable executioner of legionaries, he now joined the Legion as a private soldier.

Thirteen years later he had become lieutenant-colonel, second-in-command of the 2nd Foreign Legion Regiment at Magenta.

Sebastopol – Magenta

1854–6

On the eve of the plebiscite which was to make him emperor, Napoleon III proclaimed: 'The Empire means peace.' Perhaps there were one or two naive souls in France who believed these electoral promises. However this may be, no echo of these historic phrases – and there were plenty of them during the Second Empire – penetrated as far as Sidi-bel-Abbès where the Foreign Legion had taken root.

After the Spanish campaign the Foreign Legion was disbanded on orders from King Louis-Philippe – it was better to erase a memory painful to everybody. In 1840 it was reconstituted under the same title and raised to a strength of two regiments, the 1st and 2nd.

Though diverted from time to time by the requirements of 'pacification' in Algeria, the regiments of the Legion found the time to construct a modern model city in the marshes of the plain south of Oran. It lay at the foot of a Moslem tomb enclosing the mortal remains of the venerated Sidi-bel-Abbès. In 1831 it was no more than a skeleton camp of tents and hurriedly assembled 'boxes' – it was known as 'biscuit-town' rather than 'shanty-town'. But Sidi-bel-Abbès became the mother convent, the cradle and the refuge, the cloister and the symbol.

Patiently the legionaries and their officers constructed the barracks and in twenty years the walls had weathered, the buildings had become familiar and customs had become traditions. Among the first of these was discipline, raised to the *n*th degree until it became a sort of permanent second nature. Nowhere has the motto of the Jesuits, 'obedience unto death' (*perinde ac cadaver*) been so strictly observed as in the Legion. Complex, almost ecclesiastical, ceremonial was the accompaniment of every aspect of daily life. At Sidi-bel-Abbès during these thirty years this discipline, based on the cult of antiquity, was to leave its mark on some 340,000 recruits who passed through the place.

Here and at the training centres set up somewhat later at Saida and Mascara, during his first two months the personality of the recruit was licked into shape by the bawling of NCOs, the lynchpin of the Legion and the guardians of its tradition, by strenuous exercises in the field, by hard-fought games and by long marches. Thus he became an élite soldier, capable of enduring anything and accepting anything, even the supreme sacrifice.

But in addition to this training, which has been described as 'taming', the legionaries also acquired the pride of membership of a unique body of men. Conscious of the solidarity of his community – the 'Legion spirit' is equivalent to pride of caste – the legionary looks down on the rest of the world.

At 6 p.m. each day before the great iron gate of the barracks in Sidi-bel-Abbès a little ceremony took place and it never varied for a hundred and thirty years. The men with leave-out passes arrived at the police post – dark blue *képi* (white after 1921), khaki greatcoat or tunic. They halted in front of the great mirror fastened to a corner of the building and looked themselves over with a critical eye. If nothing seemed amiss they would then advance six paces, stand to attention and announce in a stentorian voice: 'Legionary Schmidt [or Durand or Brown]. Army number 22456.'

The sergeant did not reply; he merely surveyed the man from his boots to his *képi*. Inspection lasted a full minute and not a detail was omitted. Sometimes a toss of the head sent the legionary back; it was for him to find out what was wrong. Alternatively, returning the man's salute, the sergeant would examine his leave pass.

In the street his mates would be waiting for him a few steps away and then in little groups the legionaries would disperse into the town. It was a constricted town with streets at right angles, neither picturesque nor original, a town designed for the army and living off the Legion. There were a few Frenchmen and a lot of Spaniards – it was only fifty miles to Oran. Then, of course, there were the Arabs, confined to the suburbs or around the 'native' quarter – an exotic euphemism for the red-light district. There were one or two bars and a number of cake-shops. A tour of the town took only a few minutes.

Then the bars would fill, particularly on pay-days which came round twice a month, on the 15th and the 30th. 'Theatre' leave ended at 10 p.m. and then in their little groups the legionaries would return to barracks.

Some of them would arrive reeling a full twenty minutes before the magic hour, so drunk that they could barely stand. They would cling, swaying, to the great gate a few steps from the police post, but the sentry would not see them – it was not his duty to do so. Inside the courtyard, ten yards from the entrance, stood the sergeant of the guard, hands behind his back, facing the gate.

Outside, the drunken legionary would take a grip on himself. He had to go in and give the appearance of being sober; those ten yards between him and the sergeant would be moments of truth.

The legionary would draw himself up and then move forward stiffly – too stiffly – with jaws clenched. He would march in a straight line, almost too straight, chest thrown out, holding his breath. Ten yards . . . eight . . . five. His arm would go up in a salute and he would continue to march, eye fixed, seeing nothing. He had to get past. He was past. The sergeant returned his salute. A yard further on the legionary would collapse – dead drunk. His mates carried him off.

The sergeant would still be looking at the gate. He never turned round. As far as he was concerned the legionary had returned at the correct time and properly dressed. What happened afterwards was nothing to do with him. Drunkenness was not punished in the Legion unless it interfered with a man's duties. But woe to the legionary who collapsed during those ten yards between the gate and the sergeant! He went straight to prison.

Drink was another tradition of the Legion and a tradition cheerfully accepted. The Legion's reputation for drink has become a legend. Nevertheless the Legion did drink – a great deal . . . vastly. Every regiment has its stories of monstrous drinking bouts, of colossal beat-ups terrifying towns and alerting garrisons.

For the knowledgeable on matters concerning the Legion there is a reason for this unquenchable thirst – 'dust', the dust collected in the throats of the Legion during all its campaigns in hot countries from Mexico to Bir Hakeim, including the Sahara, Morocco and Syria. It has given rise to a tradition: a guest night in a regimental mess, when all officers are present, invariably starts with 'the dust'; with a mugful of red wine in hand, 'Le Boudin' (the Legion's marching song) is sung and then on the words 'The Dust', it is bottoms-up.

Though today beer is a competitor, for a long time red wine was the legionary's standard drink. Georges Blond, one of the Legion's historians, says that 'the quantity of red wine swilled down by the legionaries at Sidi-bel-Abbès or elsewhere would be enough to float several mighty ships'.

This is true, but it is also true that thirst is one of the worst tortures; at Camerone, at Messifré or at Bir Hakeim it was a worse agony than the fighting. In Morocco more casualties were suffered in battles for water points than in pacification. At El Mungar legionaries licked the blood from their wounds before drinking their own urine. In 1961, during the final stages of the war in Algeria, the scenes of ambushes, the points where an ambush was a certainty, were littered with empty bottles, reminders of six years of static warfare.

The legionaries made the acquaintance of thirst at Sebastopol, a name which means little to the majority of Frenchmen. In fact there is not much to say about it except that, once more, the Foreign Legion was there.

It all started one day in July 1854. As for any campaign, two mobile regiments were formed in Sidi-bel-Abbès, Russians, of course, being barred since France and Britain had decided to try conclusions with the Tsar's armies on the ludicrous pretext of the destruction of the Turkish fleet off Sinope. In fact Britain had no wish to see Russian battleships flaunting their guns in the Mediterranean.

In July 1854 the Foreign Legion disembarked at Gallipoli. From their transports the men had tried in vain to catch a glimpse of the Russians, but they were still far away and apparently in no great hurry to fight. Never mind, the legionaries thought; we'll wait.

And wait they did, all the more so since Marshal de Saint-Arnaud, commanding the operations, considered that his troops were not ready.

'If nothing happens,' the officers said, 'the legionaries will get bored' – and, as everyone knows, when the Legion gets bored, it does some pretty damn-fool things. The legionaries were not bored for long, however; they caught cholera (78 per cent of casualties in the Crimea were due to this disease).

Two months later the French took the field. On 20 September they came in sight of the Russians, 40,000 men entrenched on the far bank of a river which was to become famous and give its name to one of the bridges of Paris – the Alma.

The military manuals classify this as an extraordinary victory, but the Legion is modest and does not cite this engagement among its battle honours. It was a strange affair. On one side Menshikov's 40,000 Russians refused to leave their trenches; on the other, to the sound of bagpipes and trumpets, Lord Raglan's British army, which had not fought since 1815, advanced in faultless line as if on parade. In the centre the French, taking their cue from the Algerian wars where victory depended on the fury of the assault, rushed upon the enemy lines, their units uncoordinated and in total disorder. The Legion went in and the Zouaves went in. It was a desperate race to see who could get there first.

Once having crossed the river, complete anarchy reigned in the French ranks; carried away by the impetuosity of the Zouaves, everyone strove to reach Menshikov with no order or manoeuvre in mind. In consternation General Canrobert, an orderly person who loathed confusion, scoured the battlefield looking for anything like a complete regiment to which he could issue orders.

He spied the Foreign Legion and heaved a sigh of relief; the ranks were still visible and impetuosity had not destroyed discipline. The subalterns were still in command of their platoons, the captains of their companies and even the generals of their brigades.

'Good!' Canrobert breathed. 'Now set the others an example, my good legionaries.'

Fate is unjust, however; it is the Zouaves whom the French remember at the battle of the Alma.

Shortly afterwards the terrible siege of Sebastopol began. Like the battle of the Alma, the way in which it was conducted is still a subject of debate today. Moreover it was not a genuine siege; the flow of reinforcements to the Russian garrison was never interrupted and the gap in the Anglo-French dispositions was so wide that whole Russians divisions could pass through it to take the British in rear.

Winter came; the legionaries faced the cold and snow without complaint. They wore long 'Crimean' greatcoats, a sort of cape thrown over the shoulders. Round the stomach they bound a large cartridge-belt which earned them the nickname of 'leather-bellies' from the Russians. They were specialists in field fortifications and they knew how to dig. Undoubtedly the two regiments of the Foreign Legion were responsible for many of the fifty miles of trenches which constituted the besiegers' positions.

The legionaries soon became tired of digging, however. They wanted to fight, but they were the only people who did. When Marshal de Saint-Arnaud died of cholera, Canrobert took his place, but he became discouraged and gave way to General Pélissier. On the British side things were going no better; Lord Raglan caught cholera in his turn and died in April.

On 1 May the Legion emerged from its foxholes to take part in the offensive; it had been decided to 'degum the war'. No result was achieved and the Legion suffered casualties for nothing. Among them was one of its commanders, Colonel Viénot, after whom the Sidi-bel-Abbès barracks were later named.

Finally it was decided to launch a general attack on 18 June and capture the Malakoff, the key to the Russian defences. Six thousand British and French soldiers died in this offensive, which failed.

July went by and August came. The legionaries became bored, particularly since they were hot and thirsty. Like a paradise lost, they dreamed of Algeria and the cabarets of Bel-Abbès.

Finally on 8 September another assault was made, but in this the Legion did not take part. It was merely required to provide a hundred volunteers to precede the assault carrying scaling-ladders and planks, a thankless and particularly dangerous task but nevertheless a vital one. When the assaulting troops moved forward against the Malakoff tower they found the platforms and ladders well and truly in place with Sergeant Valliez, commanding the hundred volunteers, in the middle of them.

'After you,' this valiant NCO probably said, though history has only recorded a phrase used by a much more senior officer, MacMahon: 'Here I am and here I stay' – a typically soldierly comment which, for some reason, its author always denied that he made.

That evening the French were able to announce to Lord Cardigan, who had succeeded Raglan, that the place had been taken. The Russians evacuated Sebastopol during the night.

The Foreign Legion was the first unit into the town and the men made up, not for the fighting of which they had been deprived, but for the hardships which they had suffered. Never has a town been so thoroughly looted. The cellars of Sebastopol were full, and the legionaries, after months in the mud of the trenches under shell- and rifle-fire, could hardly believe their eyes.

They were seized by a sort of madness. The houses were packed with everything of which they had been destitute and of things which they would never possess – furniture, trinkets, sheets, linen and so forth. It was all hurled out into the streets; all was ransacked and destroyed in a frenzy of pillage.

There was not a single Russian in the town, civilian or military. No matter. The streets were full of unrecognisable legionaries, bearded to the eyes, *képi* on head, but wearing flounced dresses and carrying parasols or handbags. This was merrymaking in the raw. They looted, they destroyed anything which they could not take away, they emptied the houses. Furniture, bedding, silver, doors and windows – everything was hurled into the street, followed by trunks, trinkets, whole libraries and ending with the cathedral clock.

The final stage was reached with the discovery of Sebastopol's cellars. Russians like to drink and they had prepared for a long siege. But the men of the Foreign Legion drink even more and the communal 'beat-up' of Sebastopol is inscribed among the legends of the Legion as the name of the battle is embroidered on its colours.

The Crimean campaign ended in August 1856 and once again the Legion was disbanded. To allay the men's indignation, Napoleon III authorised the naturalisation of all foreign legionaries, though whether this was done as a gesture of goodwill or with the purpose of distributing these élite soldiers throughout the regular army it is impossible to say.

In any case the legionaries remained unruffled; on repatriation to France they simply went back to Sidi-bel-Abbès where they re-enlisted in one of the two Legions formed on 17 January 1855 to replace the regiments despatched to the Black Sea. A peculiar feature about these two Legions is that they were formed one in Dijon and the other in Besançon. These two towns had been deliberately chosen because of their proximity to Switzerland in the hope of attracting numbers of Swiss volunteers. The hope proved vain; a year later the Legions were barely at one-third of their strength.

Accordingly on 16 April Napoleon disbanded these two Legions, retaining only one, the 1st, which comprised two regiments. So the Foreign Legion was back where it started.

This was the period, however, when Foreign Legion units adopted the colour green – green tunic, crimson trousers. Green and red became the Legion's colours, which henceforth it was to retain.

The year 1857 was spent in Algeria. Revolt broke out in the Kabylie and the Foreign Legion regiments were sent to quell it under the orders of General MacMahon. With them went the Zouaves and Chasseurs with whom they had fought at the Malakoff.

The campaign is noteworthy for one engagement which is among the Legion's battle honours, the assault on 24 June 1857 on the camp of the Caid of Ichéridène. It was an operation which broke all the rules of tactics, an assault with the bayonet, up a steep slope, without firing a shot. If MacMahon realised this, he made no historic pronouncement or alternatively no one was with him that day to record it.

In 1859 came the turn of Italy. At the time when Napoleon III needed the help of the Carbonari who were fighting for Italian unification, he had promised them to reunite their country. He hesitated for some time, not because of his awkward pre-Empire profession of faith, but because he was afraid of offending the Austrians. As a face-saver he was later to send his soldiers to Mexico to support the lost cause of Franz-Josef's brother.

For the Italian campaign the Legion rejoined its habitual commander MacMahon and on this occasion he did utter one of his celebrated comments.

The French were moving on Magenta through vineyards and gardens – close country which would have deterred even a steeplechase horse. In the lead was the 2nd Foreign Legion commanded by Colonel de Chabrière with Martinez as second-in-command; on its left was the 1st Foreign Legion commanded by Colonel de Brayer.

On 4 June 1859 Magenta was in sight – and so were the Austrians. The Legion was in position level with a brick-kiln on the road from Marcallo to Magenta, its left flank covered by the Chasseurs à cheval (light cavalry). The cavalry fell back before an Austrian attack and lines of white uniforms appeared, firing volleys. For a moment Colonel de Chabrière, mounted and in his gold-braided tunic with gold epaulettes, hesitated. Then, as if speaking to himself, he said: 'No! There'll be no retreat.'

Raising his voice, he shouted: 'Packs off! Forward, the Legion!' Lowering their bayonets, the legionaries charged and the Austrians fell back, taken by surprise. In some confusion the pursuit began. 'On, on!' the legionaries yelled.

De Chabrière was astounded by his success, but, as he moved to the front to re-form his units, he was hit in the forehead and fell dead. De Brayer assumed command of both regiments and with them entered Magenta.

On receiving the news, MacMahon, with his innate sense of the apt phrase, cried: 'The Legion's in Magenta. The affair's in the bag!'

It was indeed in the bag. MacMahon became Duke of Magenta.

The next day, 5 June, the Legion buried its dead and counted its wounded. They totalled 308, including fourteen officers. On 8 June the Legion paraded through Milan.

On its colours the Legion merely carries the name 'Magenta', but on the staff of the flag is the gold medal presented to it by the city.

Camerone

30 April 1863

'What is the meaning of this?' Colonel Butet, commanding the 2nd Foreign Legion Regiment, was taken aback. With outstretched finger he pointed to an immense placard fixed to one of the windows of the Legion's barracks.

It was 15 August 1862 and the Second Empire was ten years old. It was convinced of its permanency and sure of itself, so, wherever it held sway, splendid ceremonies commemorated the birthday of Napoleon I.

At Sidi-bel-Abbès the Foreign Legion outdid all other units of the army in the magnificence and ingenuity of its celebrations. A large crowd was assembled in the great courtyard, primarily a military one, of course, but civilians were also there – civil servants or settlers. Great placards affixed to the windows told the stories of past campaigns. The Legion was only thirty years old but it had fought a great deal. Nevertheless it was the last placard which drew everyone's attention, for it was blank.

Colonel Butet was well named ('*buter*' means to dig in one's toes). He was an obstinate man, said to be capable of thinking of only one thing at a time. His short imperial beard was quivering with impatience.

The four 2nd-lieutenants who formed his escort were waiting for this question and replied in unison: 'We are keeping it for the Mexican campaign.' Then from every corner of the great courtyard came a vast shout: 'To Mexico!'

For three years the army of Imperial France had been fighting in Mexico in order to substitute for Juarez' republic the Archduke Maximilian of Austria, brother of the Emperor

Franz-Josef. But the Legion had not been included in the expedition. It was not only sad; it was bitter, for the Paris bureaucrats had selected the Zouaves of Algiers for this 'exotic' war – the baggy-trousered Zouaves. They were slightly despised rivals and in the mess it was forbidden to refer to them without adding an apologetic 'Saving your presence'. The Legion did not talk of Zouaves but of 'ouaves', on the fallacious pretext that one could not have anything to do with people like that.

Mexico! The Legion was ready to do anything, even something contrary to regulations, in order to go there. So the officers of the regiment decided to write to the Emperor, bypassing the official hierarchy. They wanted to go, they said, and they swore to make themselves useful. The Emperor sent the request on to his minister, Marshal Vaillant. He reacted according to the military book and placed the most senior officer of each rank under arrest. Nevertheless the idea began to gain ground and finally on 19 January 1863 orders were issued to form two mobile battalions of the Legion. Under command of Colonel Jeanningros, a genial giant who had begun his career as a boy in the 66th Regiment of the Line, these two battalions were to go to Mexico to 'hunt the feathered serpent'. The Legion was in high spirits. It looked upon the scene of the fighting as a treasure island and war was its secret.

The two battalions comprised seven companies together with a 'headquarters company' (responsible for supply and administration) and a band – 2000 men in all, forming a brigade. The legionaries were armed with the 1859 rifled carbine known as the 'keeper's carbine' (or the poacher's, depending on whether you prefer to be a keeper or a poacher). Its regulation title was the Minié carbine; it was a percussion weapon firing a Wessler bullet.

The 2nd Battalion was commanded by Major Munier and the 1st by Major Regnault. They sailed from Mers el Kébir on the morning of 9 February 1863, the 1st Battalion on board the *Saint-Louis* and the 2nd on board the *Wagram.* They landed at Vera Cruz on 28 March.

Mexico is a tragic country. Cortés landed there on Good Friday and this is why the slaughterer of the Aztecs felt obliged to christen the first few square yards of sand which he touched 'True Cross'.

Mexico is a rough country. The population did not like the French army and the climate was murderous; units were decimated by yellow fever, the *vomito negro.*

These unpleasantnesses hardly affected the morale of the Legion. Every man was convinced that they were about to march on Puebla, the fortress barring the road to Mexico City which had been vainly besieged for three months by the French Expeditionary Corps, 28,000 men under General Forey.

On 30 March came a disappointment. Colonel Jeanningros was told that his battalions were to be responsible for the security of the Puebla road between Vera Cruz and the little town of Chiquihuite. This was one of the unhealthiest parts of the 'hot countries', where for five months in the year the *vomito negro* raged.

On 1 April the Legion entrained at Vera Cruz. After only five miles it detrained at La Tejera where the railway still ended despite orders from the Emperor that construction was to proceed at full speed. Wearing broad-brimmed straw hats and with their packs carried by Indians and mules, the legionaries continued their way on foot towards Soledad.

As he advanced Colonel Jeanningros left detachments behind, static posts responsible for keeping watch and protecting the road. After several days marching through unhealthy devastated country the Legion reached Chiquihuite, a small town buried in the shade of enormous creeper-covered trees and watered by the Atoyac, a torrent of undrinkable water with a high sulphate-of-copper content. There Jeanningros set up his headquarters.

The Mexicans were not impressed by the new arrivals; they did not yet know about them. On 18 and 20 April Antonio

Diaz, the mayor of Jalapa, attacked the railway construction depot at the head of a considerable force. On the first occasion his raid was highly successful; the depot was sacked, the labour force scattered and the escort massacred. On the second occasion Antonio Diaz found himself facing Major Munier, governor of Tajana, and a company of the 2nd Battalion Foreign Legion under Captain Dubosq. The Legion stood to and counter-attacked at once. The Mexicans were cut to pieces and Lieutenant Milson, a Prussian officer by extraction who later served in the German army in 1870, killed Diaz with his own hand.

The Mexicans were taken by surprise. They realised that henceforth they would have to reckon with the Legion.

On 29 April a message reached Colonel Jeanningros at his headquarters in Chiquihuite. The staff of the French forces informed him that a major convoy of siege artillery *en route* to General Forey for the reduction of Puebla had been assembled at Soledad. In addition to the guns the convoy included sixty wagons and one hundred and fifty mules carrying three million gold francs – a veritable fortune. It comprised the entire French contribution to the expenses of the Mexican campaign.

Naturally this movement was top secret. Equally naturally the Mexicans knew about it. If the convoy got through, that would be the end of Puebla. If it was destroyed, it would be a very severe setback for the French. Accordingly the Mexican Colonel Francesco de Paula-Milan had collected a considerable force in the vicinity of the road from Soledad to Puebla in order to attack the convoy.

Colonel Francesco de Paula-Milan was an expert and distinguished soldier with a Greek profile, a thin moustache and the piercing eye of an artist. He was chivalrous, an hidalgo, civil and military governor 'for the enemy' of the State of Vera Cruz. To prepare his ambush he established himself at La

Joya, a few miles north of the Puebla road axis. This road climbs slowly from the Atlantic to the plateaux where the Aztec civilisation flourished in old days.

The Mexican troops consisted of about two thousand men: five hundred cavalry of the 'liberal' army, three hundred and fifty mounted guerrillas and three battalions of infantry of four hundred men each. The battalions were named after the provinces from which they came: Vera Cruz, Córdoba and Jalapa.

The Legion's task was clearly of the utmost importance. Although French headquarters knew nothing of the enemy's preparations, it considered that the outcome of the campaign depended on the security of the road. Yet Colonel Jeanningros had only two companies of the 1st Battalion available; the other two had been detached as escort to the convoy.

'One company will clear the road,' Jeanningros ordered. 'Its task will be to reconnoitre the surrounding area to discover any possible enemy positions. The second company will remain at Chiquihuite on immediate stand-by alert. Which is the duty company today?'

'No. 3,' came the answer from Danjou, his second-in-command and adjutant of the 1st Battalion.

'Call the officers. I will give them their orders in detail.'

Danjou shrugged. 'None of the officers of 3 Company is available; they are all down with fever. But no matter; I suggest that I take command.'

'Agreed.'

The officer strength of the company was made up by two volunteers from other companies, 2nd-Lieutenants Maudet and Vilain.

Captain Danjou was the son of a hosier from Chalabre, a village in Aude; at thirty-five he was a veteran of the Crimea and of Italy. Just ten years before he had lost his right hand as the result of an accident on the range in Algeria. He had had a wooden hand made to take its place which he maintained was as good as a 'real one'. He had been awarded the Legion of Honour in the Crimea. Danjou was the sort of officer the

Legion liked – broad-shouldered, always erect, high forehead, growing a little bald. He had eyes of steel.

Second-Lieutenant Maudet had been promoted from the ranks. He was thirty-four years old and had been an officer for three months. He had started life in publishing but had joined the Legion after the revolutionary period of 1848 owing to 'temperamental incompatibility with the Second Republic'. He was a brilliant soldier, and as the most bemedalled officer in the battalion he carried the colours on parade.

Vilain was the youngest – twenty-seven, though he looked even younger. He was the son of an officer, a graduate of La Fleche,[1] and had enlisted in the Legion at the age of eighteen. Although he had only been an officer for four months, he too was a chevalier of the Legion of Honour. He was boyish-looking, with curly blond hair, and was temporarily acting as paymaster.

No. 3 Company of the 1st Battalion could only muster sixty-two men out of a theoretical strength of a hundred; the *vomito negro* had claimed the rest. The oldest of the legionaries was aged just forty, the youngest eighteen. Practically every nation in Europe was represented.

The company paraded at 12.30 a.m. on 30 April 1863. Its orders were: 'Scour the country as far as Palo Verde (twelve miles to the west). Then we will push out a reconnaissance two miles beyond. We have to ensure that the road is clear and that there is not a single enemy detachment in the area. Return as soon as this mission has been completed.'

The company moved off. It was 1 a.m. The humidity was high and the atmosphere moist. The 'Hot Lands' were well named.

At 2 a.m. Captain Danjou and his men took their thirty-minutes' regulation halt at Paso del Macho, where the grenadiers of Captain Saussier's company were established.

'Are those men all you have?' Saussier asked Danjou.

1. A French boys' school which specialises in training for a career in the armed forces.

'Yes,' Danjou replied.

'I'll give you another platoon,' Saussier suggested. 'You never know.'

Danjou refused. 'Out of the question. It wouldn't give me much more strength and might risk weakening you.'

No. 3 Company went on its way. By 5 a.m. it was light.

At 7 a.m. Captain Danjou and his men reached Palo Verde, the furthest point of his reconnaissance.

'Cavalry behind us.'

'I know.'

In fact, ever since leaving Chiquihuite, Danjou had noticed that a troop of cavalry was following No. 3 Company. As he left only the horse-hoofs could be heard, but now it was daylight he could see them. They were keeping to the high ground and merely observing the French movements, barely taking the trouble to conceal themselves in the woods, thickets or orange groves bordering the road.

Before reaching Palo Verde Danjou had passed the small, semi-ruined and deserted village of Camerone. He had moved with a platoon out on either flank and they had searched the area thoroughly – 'Nothing to report'.

Just before Palo Verde Danjou, as instructed, pushed out a reconnaissance northwards towards La Joya. This was where Colonel Milan's two thousand men lay hidden. For a moment the Mexican feared that the French would find him but the legionaries did not get quite that far. He breathed again and despatched some cavalry to keep closer watch on the little force.

Meanwhile Danjou ordered a halt to 'brew up'. The weather was fine and already hot. In the Palo Verde clearing Captain Danjou's men brewed their coffee. A little cool clear stream ran a few yards away.

One section took up position as camp guard. Security was the first consideration. Rifles were piled and the mules carrying the food and ammunition placidly munched a bit of grass at the end of their tethers.

'Stand to!' It was 7.30 a.m. The corporal on duty by the stream had observed a cloud of dust rising from the Chiquihuite road – enemy cavalry.

It was the Cotaxla squadron, 250 Mexicans under the orders of Don Hilario Osario – great sombreros, long lances. Danjou recognised them in the distance. 'They are not there by chance,' he thought.

The legionaries never got their coffee. In a flash the pots were overturned and strapped back on the pack-saddles. The outpost section came in at the double and the company was back on the road without even having had time to refill its water-bottles.

Danjou's idea was simple. If the Mexicans decided to attack, he had no hope of escape in open country. On the other hand, the buildings of Camerone, 3000 yards to the rear, afforded comparatively good protection. In any case there was nothing better in the vicinity.

No. 3 Company began to withdraw in column of platoons. To prevent possible attack by the cavalry it moved along the north side of the road where thick scrub would impede the horses. This was a sound move but it threw Colonel Milan into panic. He thought that the legionaries were about to push out a deeper reconnaissance and discover him, something which he had to avoid at all costs. No. 3 Company had become a dangerous observer and had condemned itself to death.

The company had hardly reached the village of Camerone when the silhouettes of Mexican cavalry became visible on its left. They were drawn up in battle order and Danjou noticed that they had donned the heavy leather jackets which did duty as breastplates. They were going to attack.

They did attack. Danjou gave orders to 'Form square', a well-known military tactic, and in flat open country the best formation. The men stood shoulder to shoulder, rifles levelled ready to fire a volley, bayonets glistening.

The Mexicans charged, shouting, '*Viva Mexico! Viva*

Juarez!' When they were sixty paces away Danjou gave the order to fire. The charge was stopped dead. The cavalry dissolved into a mêlée of neighing, rearing horses, imprecations, smoke and dust. Having gained themselves this respite, the legionaries withdrew towards Camerone through the tangle of enemy cavalrymen and their mounts.

But Colonel Francesco Milan re-formed his cavalry on the hilltop and charged again. Danjou had transferred his company to the south side of the road where he thought that a steep bank would slow the horses down. He was wrong; the Mexican horses were as agile as goats and the obstacle made no difference to their pace. He had to form square again. A volley as intense as the first again brought the yelling charge to a halt. Then the Legion went in in its turn, counter-attacking with the bayonet to the cry of '*Vive l'Empereur*'.

No. 3 Company had freed itself and it now took cover in a large rectangular building, one side of which formed the southern border of the Puebla road. It was the main building of the Hacienda de la Trinité, separated from the hamlet of Camerone by the width of the road.

For a moment Danjou could breathe again. Temporarily at least his men were under cover.

Unfortunately, during the short battle, two of Danjou's mules had escaped and followed the horses of the retreating Mexicans. No. 3 Company now had nothing to eat, nothing to drink and only the ammunition each man carried.

The legionaries, however, believed that they were now in control of a rectangle fifty yards square. Two buildings, one on the north and one on the south, faced each other connected by two walls. Only the northern building deserved the name of house. It was a two-storey red-tiled construction bordering the road.

What Danjou did not know was that, as the company entered, one or two Mexican infantrymen had succeeded in reaching the first floor of the house and had even penetrated into the eastern room of the ground floor.

The boundary wall, built of stone and cob, was thick and about nine feet high. On its eastern side were two gates between which was a crumbling barn. Along the southern wall was another barn in slightly better shape. The legionaries established themselves in it, knocking out loopholes and organising their firing positions.

In the house at the opposite end Danjou and the remainder of the company did likewise. He barricaded the doors, blocked up a gap in the corner of the wall – through which, according to Corporal Maine, one of the survivors, one could have driven a horse and cart – and organised firing slits in the various openings. The north-western room was the best in his view, since it gave on to both the road and the courtyard. Admittedly it was a hastily organised position but at least it was better than being caught in open country.

Meanwhile Colonel Francesco Milan had called up his infantry. It advanced by forced march on the miserable hamlet where eight hundred cavalry were merely awaiting the order to reduce to pulp these sixty-five Frenchmen entrenched in the barns of La Trinité.

The cavalrymen took their time. They removed their spurs, the great rowels of which made walking difficult, and hung them on their saddles. They opened with a ceremonial fusillade to demonstrate their numbers and strength. Then they sent out Captain Ramon Laîné, son of a Frenchman who had settled in Mexico, under a flag of truce.

Laîné moved out and halted in the middle of the road. On the roof he noticed the tall figure of an observer. It was Sergeant Morzycki, a Pole, who had taken up position there to keep watch on the enemy.

Laîné addressed him in French. 'Go and tell your commander that you are lost. If you surrender, your lives will be saved. Don't wait; surrender at once.'

Morzycki jumped down from his roof and went off to report. Danjou shrugged his shoulders and rumbled: 'Surrender? And leave them free to deal with the convoy? Out of the question.'

In high spirits Morzycki climbed back on to his roof and proudly delivered the reply, adding in the best French tradition: 'Go and have your head examined!'

Laîné withdrew.

He had barely reached the shelter of a house than a hail of bullets began scoring the roughcast of the walls and chipping the stones. It continued up to the moment of the assault. The cavalrymen had taken off their spurs but their movements were hampered by the thick leather chaps designed to protect them against the spines of giant cactus. They were brave but they manoeuvred awkwardly, on the road, in the open. They were forced to withdraw leaving dead horses and men behind them.

The Mexicans persisted, attempting further attacks on the flanks from east and south. It was all in vain. Danjou had foreseen everything and his men held firm. He took advantage of a lull to make the rounds of the various posts, going from loophole to loophole. 'Hold firm,' he said to the legionaries of No. 3 Company. He encouraged, exhorted, reminding them of what was at stake.

'I shall never surrender,' he said again and again. 'That's the word of an officer.'

'None of us will surrender,' the legionaries chorused back. Together they swore a solemn oath.

Nevertheless they knew that, apart from a miracle, No. 3 Company was lost. But they were ready for anything. With their backs to the wall they would hold out as long as possible.

They were all in a highly uncomfortable position; not only had they to keep watch towards the outside to repulse the cavalrymen's charges but they also had to keep under cover from the fire directed at them by the guerrillas who had worked their way into the rooms of the house on the other side of the courtyard. The most exposed was 2nd-Lieutenant Vilain holding the tumbledown barn between the two gateways.

'Look out, Captain!' Danjou was crossing the courtyard, coming across to Vilain at the end of his inspection of the positions. It was part of his duty. He had no intention of showing fear. He crossed the courtyard erect, *képi* pulled down over his eyes, chin high and revolver in hand.

He reached Vilain, stopped, and, as if with an instinctive gesture, put his hand to his heart and fell with a bullet in the chest. Vilain rushed up and gently placed his captain's head against a stone out of sight of the enemy. But it was all in vain. Danjou was mortally wounded.

He murmured a few words among which Vilain thought that he could distinguish the word 'never'. Then he died without a sound.

Vilain took over command of No. 3 Company.

'A bugle,' Morzycki shouted, and at once the legionaries were fired by a mad hope. It was now midday and this must surely be Captain Saussier's company which had heard the firing and was coming to their assistance. It only had ten miles to cover from Paso del Macho.

'Listen,' Morzycki shouted again. 'If it's a bugle, it's an odd one.'

The legionaries listened and all hope left them. Beside the bugle could be heard the rolling of a drum, a high-pitched sound like a sputter of a basque drum, which was tapping a sort of staccato march. It was the 'degueno', the march played by Santa Anna on 6 March 1836 when the Texans were holding the Alamo, and it meant: no quarter will be given. It also meant that Colonel Milan's infantry had arrived to administer the *coup de grâce*.

There were three smart battalions, a rarity in Juarez' army, in grey linen uniforms with coloured piping. 'Vera Cruz' and 'Córdoba' wore sombreros; 'Jalapa' had a jaunty *képi* with a flap, giving it a semi-European appearance. Some irregulars accompanied the foot-soldiers – the Coscomatopec militia reinforced by peasants of the district. All were scenting blood. Most of the infantry were armed with the 1859 American

Sharp rifle, a breech-loader and the most modern weapon of the period.

Colonel Milan knew that he would win. With his 2000 men now concentrated he would butcher these obstinate Frenchmen. It would not take more than a quarter of an hour.

He felt that he could afford a little more time and display a little chivalry, so he sent back the good Captain Laîné to repeat his previous offer.

Morzycki stood up on his roof and welcomed him with a sweeping gesture. Laîné did his duty and transmitted the proposition, supporting it with a fresh argument: he pointed to the infantry drawn up ready for battle. It was in fact an impressive sight. But Morzycki did not even take the trouble to pass the message on to Vilain. 'Crap!' he replied.

Then, leaning down, he called: 'Lieutenant, I've given him the answer.'

'I know. I heard,' Vilain replied with a smile.

Laîné departed, perhaps a little sadder and certainly more lonely.

It was now 2 p.m. The five legionaries in the north-west room of the 'house' were forced to evacuate; the Mexicans had bored a hole in the ceiling and were shooting at them point-blank; they joined 2nd-Lieutenant Vilain near the barn between the two gateways.

The assault began. Vilain, who was fighting in the breach, was hit by a bullet full in the forehead. He crumpled and fell slowly. Five legionaries took his place, facing the attackers. Command of the company therefore devolved on 2nd-Lieutenant Maudet.

At 4 p.m. the fight was still raging. Waves of Mexicans broke against the legionaries' defences. At first the Mexicans were surprised by this stubborn resistance, then it exasperated them. Francesco Milan had an idea; he would smoke the French out.

The wind was from the east and on this side the Mexican infantry piled vast quantities of straw which they then set alight. The heat, which had long been terrible under a pitiless sun, became even greater; the defenders' thirst (they had drunk nothing since leaving Palo Verde that morning) increased. The barns of La Trinité disappeared in wreaths of smoke but the legionaries still held firm. The firing was as intense as ever, but when the straw had burnt out barely thirty legionaries were still fighting behind the blackened walls.

By 5.30 p.m. Colonel Francesco Milan was gnawing his fingernails. This prolonged enemy resistance was making him look ridiculous. The conflagration had changed nothing, nor had the nine-foot-wide gap opened with crowbars in the eastern wall, from which the legionaries at the western gates could be kept under fire. So Francesco Milan decided to order a general assault. He drew up his troops on the far side of the road near the cottages of Camerone and harangued them. The wind was in the right direction and Bartoletto, a Spanish-speaking legionary, translated for the benefit of the others: 'This must be brought to an end. There is only a handful of defenders. What will Mexico think of you if it becomes known that a few Frenchmen in a miserable barn have kept the forces of Vera Cruz at bay for an entire day?'

Inspired by this appeal to its national pride, the army of Vera Cruz stormed the breaches, scaled the walls and swarmed on to the roofs.

Down below in the courtyard there remained twelve men, twelve unwounded men headed by 2nd-Lieutenant Maudet and Sergeant Morzycki; the names of the other ten were Berg, Constantin, Maine, Bartoletto, Catteau, Gorksy, Kunassek, Leonard, Magnin and Wenzel.

Berg, Magnin, Gorsky and Kunassek were at once overwhelmed by the mass of enemy. They were disarmed and taken prisoner. Maudet and the seven survivors barricaded themselves in the barn at the south-east corner of the courtyard, turning a mass of debris and corpses into a parapet.

The first to fall was Morzycki who had emerged without a

scratch from the hail of fire directed at him on the roof; he was hit in the head by a bullet fired from a breach in the wall. He collapsed on top of Corporal Maine who felt someone leaning against him and on turning found himself face to face with the sergeant, dead but still standing.

'Morzycki's been killed,' he said to Maudet.

'It'll be our turn soon.'

Maine placed the sergeant's body against the wall and turned out his pockets looking for ammunition. He found only two rounds.

Fighting continued for nearly a further hour. By 6 p.m. there were only five men left – Maudet, Maine, Constantin, Catteau and Wenzel.[1] Each of them only had one round of ammunition left.

'We'll sally out,' Maudet ordered. 'Fire when I give the word.'

The thin little volley rang out and then, with levelled bayonets, the five men charged the army surrounding them. Dozens of rifles were lowered. The Mexicans were about to reply. At a range of fifteen feet it could only be a massacre.

Maudet led the charge. At the very moment that the Mexicans fired Catteau bounded forward and threw himself in front of his officer, covering him with his body. He was hit by nineteen bullets simultaneously. But his self-sacrifice was in vain. Maudet was hit twice and fell with his thigh broken and his hip mangled.

Then a Mexican officer from the 'Jalapa' battalion, in *képi* and with officer's stripes, moved out in front of his men and ordered the cease-fire. With his sword he struck up his men's bayonets and, turning to the legionaries, called: 'And now will you surrender!'

Maine, the most senior NCO, replied: 'I accept, on con-

1. The Legion's appraisal of the four soldiers was as follows: Maine: French: brilliant in the field; gave up his sergeant's stripes to fight in Mexico. Catteau: a 'true' Belgian with less than six months' service. Constantin: serving under a false name; headstrong but good in the field. Wenzel: an old reliable veteran.

dition that we keep our weapons and that the Legion's wounded are cared for at once.'

A colonel came up. His name was Combas and, like Laîné, he was of French extraction: 'We can refuse nothing to men like you,' he said.

The legionaries were led to the centre of the courtyard.

'Where are the others?' an officer asked in astonishment.

'That's all,' Maine threw back at him proudly.

Colonel Francesco Milan arrived. He too was astounded and summed the matter up by saying: 'They're not men; they're demons.'

As Maine and Wenzel followed by Constantin emerged from the building giving on to the courtyard accompanied by the Mexican Colonel Combas, one of the cavalry irregulars hurled himself on them yelling and brandishing two pistols which he fired point-blank at the two leading legionaries. He missed. Combas drew his revolver and shot him.

Colonel Francisco Talabera, commanding the 'Córdoba' battalion, was also a doctor. He had all the wounded carried some hundred yards to a little embankment rising north of the road.

It was 6.30 p.m. After eleven hours of fighting No. 3 Company of the 1st Battalion of the Foreign Legion had ceased to exist. Of No. 3 Company's total strength three officers and forty-nine legionaries were dead, or died shortly afterwards from their wounds. Twelve were taken prisoner, almost all wounded. Four men were never seen again.

Colonel Francesco Milan and his 'Centre' brigade were left masters of the battlefield – but at what a price! His purpose, to attack the convoy of artillery and gold, fell through since surprise was of the essence and the French were now on their guard. A few days later both guns and gold arrived intact and Puebla capitulated. By forcing the enemy out into the open and pinning him down for a whole day Danjou and his men had done more than merely fight. They had saved the entire French expeditionary force. The 3rd Company of the 1st Battalion of the Foreign Legion no longer existed but its annihilation had purchased victory.

Warned by Indians, Jeanningros concentrated his forces next day. He marched all night, gathering up Saussier's grenadiers as he passed, and arrived within sight of Camerone at dawn on 2 May.

Less than two miles from the hamlet a sort of ghost emerged from the undergrowth, a man covered in blood, unable to speak and wearing only a shirt. He made himself known by sign-language. His black, swollen tongue protruded from his mouth. The legionaries could barely recognise their comrade Lai, the drummer. He had received seven lance-thrusts and two bullets and, after having been stripped of his equipment and clothing like the rest, had been left for dead and thrown on to a pile of corpses by the Mexicans who were too indolent to bury the bodies. When night fell he regained consciousness, crawled out from under the other bodies and set off along the road to Chiquihuite. All through the night coyotes had been sniffing at him but they are hesitant to attack a live man. He thought that he was the sole survivor of No. 3 Company.

In the great courtyard of La Trinité were twenty-two bodies together with scattered remnants of uniform, equipment and weapons. Most of the bodies had been attacked by coyotes and several of the legionaries were unrecognisable, but someone found Captain Danjou's wooden hand.

Outside the bodies of ninety-six Mexicans were scattered around, also half-eaten by coyotes. Their nationality could be told from the fact that they had not been stripped of their uniform. Near the embankment where the wounded of both sides had been taken, a shallow trench held some fifty further bodies.

'Vengeance for Camerone!'

Colonel Dupin set off in pursuit. He had originally been a Foreign Legion officer but for some time had been commanding a counter-guerrilla force consisting of a collection of adventurers, bankrupt slave-traders, malcontent sailors, unlucky gamblers, despondent planters and soldiers of fortune. It was a sort of gang carrying out its own form of irregular warfare against other irregulars. As his second-in-command Dupin had selected an officer as dare-devil as himself and bearer of a famous name – Ney d'Elchingen. The uniform worn by this corps of freebooters was less stereotyped but more practical than that of the legionaries – broad-brimmed flimsy hats, baggy trousers, soft boots and wide linen cummerbund. Dupin had one hundred and fifty men under his command.

Together they intoned their war-cry: 'Vengeance for Camerone!'

Dupin heard of the massacre on 3 May. He passed through Soledad, borrowed two companies of the Legion and set off in pursuit. On 5 May he missed the Mexicans some miles from Soledad but refused to be discouraged.

Through some Indians whom he sent out as scouts he soon learnt that he was on the right track. On 13 June he raided a village said to have provided a certain number of guerrillas for the Camerone affair. Dupin's one hundred and fifty cavalry attacked the village which was named Cueva Pentada. The Mexicans fled as he arrived but he captured eight whom he shot on the spot.

Then the village was searched. Numerous items such as uniforms or *képis* belonging to No. 3 Company were found, and one wounded prisoner, Legionary de Vries. From him Dupin learnt that Cueva Pentada had been the starting point for Cotaxla's cavalry. Dupin burnt the village to the ground.

On 28 June Ney's Indian scouts warned him that enemy forces from Jalapa and Perote were on the move towards Huatusco. By 8 p.m. that evening, after a fifteen-mile forced march, Dupin with his cavalry and a hundred and twenty-three legionaries 'lent' by the 1st Battalion and commanded by Captain Sudrie and Lieutenant Perret succeeded in reaching Huatusco before the Mexicans. He left the village forthwith and went out to meet them.

At 11 p.m. Sudrie made contact with the enemy advanced guard. While Perret carried out a pincer movement with his infantry, Sudrie attacked and Dupin's yelling cavalry came up in support.

The Mexicans, who had already bivouacked, panicked and fled. To escape the French the majority dashed down into a precipitous ravine with the horrific name '*Barranca espinosa del Diablo*' (Thorny Valley of the Devil), henceforth a black name in Mexican military history. Having reached the bottom and thinking themselves still pursued, the fugitives began shooting each other. Dupin and the Legion contributed to the slaughter from the top. Thirty-five dead and forty-five wounded were found down there next day. Eleven Mexicans from Jalapa had already been killed in the fighting on the plateau.

Dupin returned to Huatusco. This was the village in which 2nd-Lieutenant Maudet had been buried with full military honours by Colonel Francesco Milan's troops. For several days he had been cared for by a great-hearted young Mexican lady, Dona Juana Marrero, Countess of Gomez, who was said to have soothed his last moments. Her two brothers, Manuel Marrero and Don José, had handed Maudet over to her.

The French did not forget and when Manuel Marrero was killed in an ambush two years later, two officers of the Legion carried the body back to Huatusco.

On 18 February 1867 the 1st Battalion of the Legion left Mexico on board the *Pomone*; on 20 February the 5th and 6th Battalions followed in the *Aveyron* and on 21 February the 2nd Battalion in the *Var*. The Legion's campaign finally ended when the 3rd and 4th Battalions left Vera Cruz on the 22nd on board the *Tarn*.

In a little under four years the Legion had lost twenty-two officers and 446 NCOs and men.

Paris

1870–1

It was raining. It had been raining uninterruptedly for four days. October 1870, still young, was drowning everything in an icy deluge. The unhappy XV Corps, hurriedly formed at Tours, was engulfed in mud as it moved forward from Orleans to relieve the siege of Paris.

In the lead was the 5th Battalion of the Foreign Legion. It was an odd unit which had never been under fire nor baked in the African sun; it consisted of volunteers and had been hastily formed at the end of August by decree signed by the Empress Eugénie. Napoleon III, with vacant eye and flabby hand, had left for Sedan where he was to immure himself with his army in an inglorious siege. A few days before he did so, General Leboeuf, who had just assured an enthusiastic Assembly that 'not a gaiter button was missing', had brought the Emperor a message from the head office of the Customs: 'Every day a considerable number of foreigners present themselves at the frontiers asking to serve under the French flag for the duration of the present war. ...'

With a weary gesture Napoleon had agreed in principle that a fresh foreign unit be formed.

Four battalions of the Legion were serving in Africa. The battalion formed at Tours was therefore numbered 5. A commander was found for it – Major Arago, son of the famous scientist. On 8 October, after four weeks' training, the 5th Battalion of the Foreign Legion moved to Orleans. On the 10th it reached Artenay, seventeen miles to the north.

The Germans were there already. To guard against a

threat to his southern flank Moltke had detached a division of 20,000 men under von der Thann from the forces investing Paris. It was composed of Prussians and Bavarians and included both infantry and cavalry. Facing them were Belgians, Serbs, Wallachians, Italians and Spaniards. They were volunteers but they were inadequately trained and poorly armed. They had a few chassepot rifles but most of their weapons were old percussion-locks dating from the Second Republic. Nevertheless these newly fledged legionaries clung to their ground following the example of one of their officers, Pierre Karageorgevich, a young Serbian officer, one day to be king of his country. They clung on while the Uhlans and Bavarians, after two abortive attacks, bombarded Artenay with their artillery. The thatched roofs were burning, the peasants crossing themselves.

'Fortunately,' said the practical-minded women, 'the invasion came after the harvest.'

Meanwhile XV Corps was at Saran in the suburbs of Orleans, seven miles away. Reluctantly Major Arago gave the order to withdraw. Though only a newly joined officer as far as the Legion was concerned – he had come straight from the 'regular army', the 86th Infantry – he disengaged 'in the manner of the Legion', after a bayonet charge.

The 5th Battalion reached Saran when XV Corps' rearguard had already crossed the Loire and was preparing to fight far back, at Tours.

Gambetta, however, who had escaped from Paris in a balloon, tried to rally the failing courage of the military and cabled to General (Reserve) de la Motteraye commanding XV Corps: 'No question of retreat; defend Orleans.'

For a time this order was obeyed. The Prussian advanced guard reached the suburb of des Aydes where hand-to-hand fighting took place with the 5th Battalion which only gave way slowly. The ground, moreover, favoured the French, as the fighting surged through small gardens divided by hedges and low stone walls.

Von der Thann was infuriated by this further resistance. He was dreaming of parading through Paris and feared that the siege of the capital would be over without him. At the risk of mowing down his own infantry he set his artillery to work. As at Artenay fires started.

Fearing that the Germans would outflank him from the west – ever since Sedan the French army had been obsessed by a threat of encirclement – la Motteraye decided to disregard his orders and evacuate the city. XV Corps began to move on 11 October. Certain units were left behind to cover the withdrawal and among them, of course, was the 5th Battalion of the Foreign Legion.

It clung on still, though forced to withdraw. The Germans loosed a hail of fire; their infantry dashed forward as if out of the mouths of their guns. The French were so close that they shot the Germans down as soon as they appeared through the smoke. Still leading his battalion was Major Arago, stick in hand, cigarette in mouth, encouraging his men, directing the volleys himself. He was killed at 5 p.m. on 11 October.

But the 5th Battalion still fought on. At 7.30 p.m. it was holding the edge of the suburb of Bannier. Just at this moment firing was heard from the rear. Without warning XV Corps had disengaged, leaving the way open to the Bavarian infantry which drove into the gap, cutting off the legionaries' retreat. They were called upon to surrender but refused. They had to be winkled out of houses and cellars or shot down in gardens. It took the Bavarians all night by the light of the burning buildings.

By the morning of 12 October the 5th Battalion of the Foreign Legion had ceased to exist. In forty-eight hours of fighting it had lost two-thirds of its strength. It earned the admiration of the enemy. General von der Thann said: 'If the French had fought like this at Sedan, we should never have reached Orleans.'

Nevertheless a few officers and legionaries succeeded in escaping from the trap in little groups, most of them in civilian clothes, having been hidden by the inhabitants of Orleans. Pierre Karageorgevich succeeded in rejoining XV Corps at Pierrefitte disguised as a miller.

That same day the Foreign Legion Regiment from Africa, which had not taken part in the war, disembarked at Toulon. It was to be involved in the final stages of the fighting, to save the honour of France.

The order to sail had not been received until 6 October, a month after Sedan. The officers of the 1st Foreign Legion Regiment, seething with impatience, had heard with consternation of the slow disintegration of the front; if there were no victories to be won, at least they would have wished to be there to die with the rest.

Most of the African troops had left Algeria – Zouaves, Turcos and Chasseurs – and the authorities had tried to explain that soldiers were still required 'to make the Arabs feel the weight of French might'. But this was only academic argument.

Finally on 8 October two battalions of the 1st Regiment left Sidi-bel-Abbès by train. According to tradition they marched to the station behind the Foreign Legion band which consisted of Germans. Now it was up to Germans to 'maintain the French presence in Algeria'.

On 10 October the 1st Regiment of the Foreign Legion disembarked at Toulon. It found the town deserted; having heard of the arrival of the Legion, the inhabitants imagined that the Germans in its ranks would take the first opportunity to change sides. The Legion shrugged its shoulders and took no notice. On 13 October, with the rain still falling, it arrived at Bourges. Next day, at Pierrefitte, it found the survivors of the 5th Battalion, whom it absorbed.

The 1st Regiment was now in the Army of the Loire, formed by Gambetta at Salbris and commanded by General d'Aurelles de Paladines. In the space of a few days 60,000 men assembled facing von der Thann's 20,000 Germans. For the first time since the start of the war numerical superiority was on the side of the French.

On 9 November there took place the battle of Coulmiers. It was the first and only French victory. Von der Thann withdrew, and the next day he evacuated Orleans. He left as rearguard a battalion of Bavarians commanded by a colonel whom the veterans of Mexico knew well – ex-Lieutenant Milson, the hero of Jalapa.

Gambetta's idea was that regional armies, commanded by well-known officers such as Chanzy and Faidherbe, should converge on Paris. In fact, for a reason never satisfactorily explained, he ruined the whole plan by a faulty manoeuvre: he strung the Army of the Loire out over a front of forty-five miles. The error cost him the loss of Orleans for the second time.

Captain Labarrière was furious. On 8 October, the day when the Foreign Legion Regiment left, he was stuck in hospital at Géryville south of Oran with jaundice. Why risk a relapse?

Like his fellow-officers he had been grumbling for months. Here he was, condemned to inactivity while the army was fighting – and on his native soil. On numerous occasions he had asked to be posted so that he could fight for his country but each time his request had been rejected as 'purposeless'. When he missed the sole opportunity of campaigning in France, this was for him the height of misfortune.

Two days after the regiment had left, he took a decision: 'Since there is no authorised way of going to war, we will go and fight without permission.'

Knowing the ways of the impetuous captain, the medical officer had taken the precaution of hiding his uniform. His only clothing was a white woollen jacket made from a hospital blanket. 'Never mind!' Labarrière growled; 'it may make me look like a pensioner but I'll use it as a cape.'

By 15 October he was wandering round the port of Oran looking for a boat. That evening, with the help of some chance friends, he embarked as a stowaway. The following evening he arrived in Marseille.

He had forgotten, however, that in wartime the military police were insistent on regulations. He had neither papers nor movement order and he could give no plausible explanation for his extraordinary uniform, on which he had sewed officer's stripes. Labarrière was carted off to prison, accused of a serious crime for an officer – desertion in wartime.

Fortunately the Legion's guardian angel, in this case one of his fellow-officers in the Army of the Loire, heard of Labarrière's unhappy predicament. Finally, on government orders, the impetuous captain was despatched to join his unit.

He arrived on 4 December; transport was not then as rapid as in our day. But it did not matter, Labarrière was happy. He leapt from the train at Orleans and dashed off to present himself to his commander, Colonel Canat.

'What's the meaning of this garb?'

'It's just an idea, Colonel. This jacket wanted to go to war.'

Canat shrugged; this was no time to discuss turn-out. Labarrière went off to join his company, which was already in the line between Cercottes and Chevilly, about eight miles north of Orleans.

The previous day, 2 December, the Army of the Loire had been defeated at Loigny. The outcome of the war was now decided. All that the Foreign Legion could do was to fight to the end.

Thick snow had been falling for two days, burying the countryside and leaving the blackened skeletons of the houses standing out against the white and grey of the sky. An atmosphere of disaster lay over everything, reminiscent of the Russian campaign. The legionaries were in rags, many without boots and with frost-bitten feet. Nevertheless they were expected to bring the enemy attack to a halt and they were to show that they could be relied upon.

The winter of 1870–1 is still famous for its severity. The frozen trees cracked like glass under the shell-bursts; roads and springs were frozen. So were the rifle barrels; if gripped with the bare hand the skin of the fingers was left behind.

Opposite were the German positions; they had not even taken the trouble to dig in – what was the good; it was a waste of time. The legionaries could see the ghostly grey-green silhouettes of the Bavarians in their round balaclavas preparing for the assault.

Far out in front, crouched behind a milestone, was Legionary Weber, an Alsatian and ex-sergeant reduced to the ranks. He had decided to stop a certain German battery firing and he shot down three of the gunners in a row.

'Watch out,' shouted Labarrière who had just come up. 'Come away from there.'

Weber obeyed reluctantly. Almost immediately two shells landed on the spot which he had just left.

'What an honour!' he chuckled. 'The Bavarian artillery making me a present of two shells.'

But Labarrière did not reply. The enemy attack had started. It emerged from two sunken roads while the artillery, firing shrapnel, prevented any counter-attack by the flanks.

Then, remembering Camerone, Labarrière ordered, 'Fix bayonets!'

From a distance of fifty yards, Bavarians and legionaries fell upon each other. It was grisly. One hundred, perhaps two hundred men bearing down, bent on tearing each other to pieces.

Labarrière was ahead of his men, still wearing his white blanket, brandishing his sword. A Bavarian aimed at him point-blank and Labarrière cut him in two. Then came a general mêlée. It lasted only a few minutes and the enemy was thrown back. Labarrière charged on in pursuit. He was killed by a bullet in the head fired from a distance by an infantryman concealed behind a hedge.

Nevertheless the impetus was maintained. Lieutenant

Ravaut, Labarrière's second-in-command, dashed forward, jumped a hedge and fell upon a group of Germans consisting of a 2nd-lieutenant and four men. He shot the officer with his revolver and put the others to flight.

Such isolated deeds of heroism, however, could not save Orleans and that night, in bitterness of soul, the regiment evacuated the city. At its head was a curious individual of whom history records nothing but the fact that he was a journalist and spent more time sending despatches to the newspapers than in commanding his unit.

Back at Chapelle-Saint-Ursin the Foreign Legion absorbed 2000 additional men from the 39th Infantry. They were Bretons and hardly spoke French. A few days earlier, when they had halted, miserable, haggard and half-dead with cold and hunger, they had been visited by General d'Aurelles de Paladines. They hardly bothered to get up. A long-drawn-out shout came from this motley crowd, dazed with fear and fatigue: '*Ar ger, Général, ar ger*'.

In Breton the 'g' is hard and therefore the word sounded like 'guerre' (war). The general mistook their meaning. 'Oh, these brave fellows,' he said. 'They want to fight for France.' No one had the courage to tell General d'Aurelles de Paladines that these Bretons, who were fighting under their national black-and-white flag, were not trying to say 'Off to war' but 'Off home'.

Nevertheless the Legion made them welcome and, in contact with the veterans of Mexico and Africa, these 'strangers from the west' did as well as their comrades, the 'strangers from the east'.

From Chappelle-Saint-Ursin the Foreign Legion was sent to Vierzon where a train was waiting. What was its destination? 'No one knows,' the officers replied. 'In any case, someone added, 'as we're taking a train, it's that much less to do on foot.' Cold comfort. The Legion did not know what the future held for them.

The train set off eastwards, desperately slowly. The cold was arctic and in Besançon station some trains were frozen in for five days in a temperature of −18°C. It was a hellish journey.

Then one night came the cry: 'Last stop! Everyone out!'

The legionaries asked each other: 'Where are we?' No one knew. Some of the more resourceful, however, managed to discover that the regiment was 'somewhere near Montbéliard'.

'Montbéliard? What for?'

From Vierzon to Montbéliard seemed a peculiar move, particularly since Prussian artillery had now been bombarding Paris for a week. One would have thought that the most urgent thing was to march on the capital. The ordinary legionary obviously could not know, however, that General Bourbaki, the sole general still with an 'army' intact (Chanzy had been defeated at Le Mans), had been ordered to break through the enemy lines from the east.

Bourbaki is a name which the legionaries who served in his army will long remember – and without affection. The spectacle presented by this heterogeneous body of troops was indescribable and its effect on morale disastrous. To this was added cold and hunger. In forty-eight hours the men had seen nothing but a hunk of bread, frozen and hard as stone, and a repulsive piece of fat. Nevertheless this was the force which, for a moment, was to give the impression that it had won a victory.

The Legion arrived on 13 January and attacked on the 14th. Facing it were 45,000 Germans under Werder, the victor of Strasbourg. Objectives were the heights of Sainte-Suzanne and the guns defending Montbéliard. The Legion attacked at dawn. By midday it had captured all its objectives and Montbéliard was in Bourbaki's hands. But every single unit was out of ammunition! There was no alternative but to abandon the town, and the Army of the East withdrew to Besançon. In fact it disintegrated and most of the men quite simply went home to escape this war which was so near its end.

The Legion alone stuck to its duty and established itself at

Besançon which it prepared to defend. Following its example, the Bretons of 39th Infantry helped in the work. Moreover the Foreign Legion Regiment gave proof of its pertinacity and confidence in the future: it continued to enlist recruits! The armistice with Germany was signed on 28 January. On 17 March the regiment reached Versailles with a strength of 66 officers, 1007 NCOs and men. Taking account of reinforcements received during the campaign, in three months the Legion had lost 14 officers, 916 NCOs and men killed.

Starting on 7 April the Foreign Legion was despatched to fight against the Paris Commune. The Legion's Golden Book published in 1931 to commemorate its centenary, says: 'This is not the duty of a legionary and we will pass over this unhappy episode in silence.' For all Frenchmen this is one of the most tragic incidents in history. Numerous atrocities were committed by both sides; the fighting certainly was not pretty. The Legion lost two officers.

From 7 to 27 April the Legion was fighting on the western outskirts of Paris, between Neuilly and Courbevoie. After a week's rest it was again in action in Neuilly. It occupied the southern part of the sector, a line inside the Bois de Boulogne.

On the morning of 5 May the outpost patrols exchanged shots with the enemy. Suddenly firing stopped, though no orders had been issued.

'What's happening?' asked the sergeant commanding the patrol.

'Look!'

The sergeant stared. A hundred yards away an extraordinary apparition was slipping through the trees; a small, black, completely naked figure galloping towards the lake. Behind it was a man in white yelling: 'Don't shoot! Don't shoot!'

Finally the legionaries realised what it was and burst out laughing. The small black man apparently causing such agitation to both sides was no less than the orang-outang from the zoo who had taken to his heels.

For another fortnight the Legion fought on the northern edge of Paris. On 27 May it stormed the Buttes-Chaumont and was the first to hoist the tricolour flag there.

The end was a strange paradox. The following month the survivors of the campaign returned to Algeria. There was the band at Sidi-bel-Abbès station. The Germans of the Legion were faithful to its traditions.

Dogba

12 September 1892

'WIDA! ALL ashore'.

The legionaries laughed. They found everything amusing in this kingdom of Dahomey with its queer names, starting with that of Ouidah, the little port where they had just disembarked in August 1892, and ending with that of the negro king who ruled the country. His name was Behanzin and his nickname was obvious – Bec-en-zinc (Zinc-nose).

He was said to be an odd personage, always clad in a striped robe. About fifty years old, with grey hair and beard, he was an authoritarian and exacting tyrant whose word was law among his population of slaves. Ouidah, the coastal port, was the main export terminal for these 'voluntary workers', the euphemism necessitated by the abolition of slavery.

Behanzin astounded the legionaries. In the first place he voiced a lofty disdain for death – that of other people of course; secondly he kept as his bodyguard a fascinating force consisting of fiercely enthusiastic female warriors whose primary occupation, apart from making war, was to carve into small pieces, in the literal sense of the words, the prisoners, the wounded and the dead.

At this moment the legionaries were disembarking with stomachs in a certain turmoil; they had had to cross the 'bar' in launches tossed hither and thither by the waves and their arrival was not particularly glorious. It bore little relation to the picture presented next day by the Paris *Petit-Journal*; this showed the Legion marching past, headed by its band, along a wharf which looked like Deauville jetty. But legionaries have always been ready to forgive the journalists their flights of fancy.

In fact the very next day the Dahomey battalion of the Legion – 800 men under Major Faurax – took the trail. Their destination was Kana, King Behanzin's capital. The French forces (3000 men in all under Colonel Dodds) were to go and convince the ruler, who had been tempted by the blandishments of the Germans, that it was better to remain on good terms with France. The army is always an excellent method of persuasion.

Major Faurax' column left on 1 September to join up with an advance detachment which was to 'subjugate' the Décamé, an area along the River Ouémé. Guided by Senegalese scouts, Faurax pushed on for ten days, covering only four to five miles a day through a forbidding forest in the depth of which nothing was visible, neither the sky nor the sun, nothing but the green roof of creepers and trees. They marched by compass and were soon hacking out their own path. After the forest came the savannah, six-foot-high grass through which the column carved a sort of a path using its swords, a path which vanished again instantly.

For ten days the column advanced in this fashion; ten days during which the legionaries forgot how to laugh. Their faces and hands were covered in mosquito bites and there was nothing to eat but army biscuit. Thirst raged – once more. The cry was: 'Get on, get to the river.' It was reached on 10 September. Faurax was a kindly and far-seeing commander. At the first village which they passed he had requisitioned porters to carry the legionaries' kit. For the first time in their history, moreover, they were suitably clad, in colonial-type tunics of khaki-coloured linen and white trousers. Each morning there was an issue of quinine to guard against malaria.

On arriving at the river the Legion found the right bank, said to be dangerous, being patrolled by gunboats of the French navy which was in overall charge of the operation.

Faurax joined up with the French column at Fauvié on the left bank.

On 11 September the whole force was on the move again. That evening it reached Dogba, a little village on the river bank where Faurax decided to camp. To the west, on the river, were the navy's gunboats. Camp was pitched on the bank in a square with sentries guarding the approaches.

The legionaries were happy. They had finally come to like this country; the Dahomean girls were pretty and the temperature acceptable. Moreover the river was there now – nothing to worry about.

'We're well set up,' they said. And so they were – against everything except an attack by thousands of fanatical warriors. But this is what happened.

At 5 a.m. on 12 September, when the bugler sounded reveille, thousands of Dahomean warriors emerged from every thicket and every bush. Some, who had climbed up into the trees, even hurled themselves on the French from above.

Like the veterans they were, the legionaries fell back but did not run. They seized their arms and were ready on the instant. The blacks were armed with lances and pangas – terrifying weapons but less effective than the chassepot rifles which tore holes in the serried ranks of the attackers.

'Forward!' Yet once more the Legion charged. Their first thought was to extricate themselves from this mass of human beings who clung to arms and legs, engulfing the legionaries, smothering them, stifling them until a man was left hacked to pieces as if by a hay-cutter. Then, with the bayonet, they began to push back the raging ranks of their enemies.

'Forward!' The shouts of the legionaries answered the yells of Behanzin's warriors. Some of them were armed with rifles, and well armed too; Germany had provided them with new Mauser rifles and ten-round Winchester repeater carbines. These men did not enter the mêlée, but fired from a distance, supporting the Dahomean assault.

Four times the Dahomeans charged. Four times the Legion

drove them back despite its losses. Fighting raged in the camp until 10 a.m. Then the blacks withdrew, discouraged. They left a thousand dead on the field including some sixty of the female warriors whom, for want of a better word, the legionaries christened 'Zinc-Nose's Amazons'. The women wore the traditional loin-cloth, held at the waist by a leather cartridge-belt; their breasts were bare and on their heads was a sort of red tarbush. They would never surrender and their cruelty soon became legendary.

On the French side the attack had cost the Legion forty-five dead. Among them was Major Faurax who, with his last breath, sent a message to Colonel Dodds: 'Have my men given satisfaction?'

Faurax having been killed, Captain Battreau, commanding No. 1 Company and the next most senior officer, took command of the battalion. The force set off once more, still aiming for Kana, where Behanzin was said to be. It was tough going, tougher still along the river owing to the vegetation and incessant attacks by Dahomeans emerging from the bush in never-ceasing waves.

On 4 October the Amazons made a further appearance. They had the good fortune to encounter the Senegalese scouts of whom they had no fear, and, in spite of resistance, drove them back. The women fought with the utmost fury, clawing at their enemies and even biting the bayonets which transfixed them.

The fighting ended at dawn and the legionaries scouring the battlefield found the explanation for this superhuman courage – the lovely Amazons were drunk as coots! At their starting point they found several barrels of tafia (a kind of rum), unfortunately empty, and numbers of gin bottles, equally empty. 'No miracle here,' said the disappointed legionaries who thought they had a potion for immortality.

By 14 October the river bank was found to be impassable

and the column had to move inland. Thirst became a problem again. There were practically no supplies and Behanzin, who was well informed, denuded the countryside ahead of the column. In order to survive, it split in two; half went ahead to open up the track while the other half went hunting. On the first occasion all went well; a herd of cattle was discovered. Four days later nothing could be found. The convoy was forced to halt to allow supplies to catch up. Two days went by.

On the morning of the third day the Legion dissolved in laughter: the 'hunting party' returned carrying stretchers. 'It's not possible,' they said. 'Have you been hunting each other?'

They took a look at the 'wounded' and the laughter became even louder. The party had found a herd of cattle but they were so thin, so small and so rickety that the men were afraid that they would collapse *en route*. 'It was better to carry them,' they explained.

On 6 November the French column entered Kana. 'Call this a capital?' the legionaries asked, disgusted by the foul grisly huts of the witch-doctors, lined with grinning masks, human skeletons and jaws of animals. 'It looks like a museum of anatomy.'

A week was spent in Kana waiting for Behanzin to surrender. Finally he decided to decamp and abandoned his last town, Abomey north of Kana. The French entered it on 16 November.

The legionaries took over the king's palace. It was solidly built and the cellars contained excellent bottles of vermouth, champagne and bordeaux. 'Conquests bring one something in the end,' the legionaries said.

They returned to Bel-Abbès with a truly royal trophy: a magnificent white parasol, the circumference of which was ornamented with fifty human lower jawbones.

'When a French soldier goes into hospital, he is eventually sent back home. When it is a native soldier, he is eventually discharged in good health. When it is a legionary, he comes in to die.'

This was the terrible comment of a hospital commandant. It summarises in a few words the Madagascar campaign and it does so without exaggeration. The figures speak for themselves: total strength of French expeditionary force – 21,600; killed in action – 7; died of wounds – 13; died from various other causes – 5736.

Once more the Legion was to distinguish itself. The French Republic had quarrelled with Queen Ranavalo III and France decided to send an expedition to 'bring her to heel'.

After Dahomey, a victorious campaign lasting two months, it was now the 'Great Island' which was to experience the weight of the French arguments. The 1870 defeat was now old history. In any case it was the Empire that had been defeated; the new Republic was pacific and liberal.

Behind General Duchesne, its commander-in-chief, a regiment of the Legion disembarked at Majunga on 6 May 1898. It was thunderstruck by the organisation; the port was chock-a-block with equipment. 'A bit different from Dahomey', the old hands said.

In fact the Supply Service and the recently organised Ordnance Service had decided to provide all units with vehicles. The principle was simple: 'How many kilos does a mule carry?' – 'About a hundred' – 'And a cart drawn by a mule?' – 'At least four times as much'. QED. The French army in Madagascar would be equipped with carts specially designed for colonial conditions. Clearly the Supply Service and the staff had provided for everything; it was a far cry from the negligence of the Imperial period.

In fact they had provided for everything – except roads.

Never mind! The expeditionary force would simply have to build one. On the map the distance between Majunga and

Tananarive, the capital, was only a hundred miles. A mere nothing.

Five thousand vehicles were assembled at Majunga. That road cost the French 5000 men – one per vehicle.

The legionaries stood up to the conditions better than the soldiers from France; they had fought all over the world and were better trained. They were the first to enter Tananarive on 26 September with General Metzinger at their head. The nightmare was over, and the expeditionary force could return to France.

Seven months later, when another force had to be despatched to 'pacify' Madagascar again, General Galliéni, who was in command, said: 'I ask to take with me six hundred men of the Foreign Legion so that, if necessary, I can die decently.'

In fact the Legion sent more than 600 men to Madagascar. For seven years, until 1905, the Legion maintained a regiment there, keeping it continuously up to strength with drafts from Sidi-bel-Abbès. The tour of duty was two years.

PART TWO

Susanna

1900–40

Susanna

Marching Song of the Mounted Companies

Ah! les carottes! Ah! les carottes
Les carottes et les navets!
Qui c'est qui s'les tape (bis)
C'est la Compagnie Montée
Oh, Susanna, les carrottes et les navets.
Oh, Susanna! Compagnie Montée!

Ah! les djebels. Ah! les djebels
Les djebels et les pitons!
Qui c'est qui s'les tape (bis)
Oh, Susanna, les djebels et les pitons!
Oh, Susanna! Compagnie Montée!

Ah! les femmes. Ah! les femmes.
Les femmes et les fiancées!
Qui c'est qui s'les tape (bis)
Oh! Susanna, les femmes et les fiancées!
Oh! Susanna! Compagnie Montée!

Oh ! The carrots ! Oh ! The carrots !
The carrots and the turnips !
Who is it who gets them (repeat)
It's the Mounted Company.
Oh, Susanna, the carrots and the turnips.
Oh, Susanna ! The Mounted Company

Oh, the desert. Oh, the desert
The desert and the bloody noses.
Who is it who gets them (repeat)
Oh, Susanna, the desert and the bloody noses.
Oh, Susanna ! The Mounted Company.

Oh, the girls ! Oh, the girls !
The girls and the fiancées !
Who is it who gets them (repeat)
Oh, Susanna ! The girls and the fiancées !
Oh, Susanna ! The Mounted Company.

El Mungar

2 September 1903

'PARA. 1: the commander is always right. Para. 2: if, exceptionally, the commander should not be right, Para. 1 becomes applicable at once.'

Lieutenant Selchauhansen was a poker-faced wag. He was tall, phlegmatic, adored by women and a first-rate tippler, just what a Foreign Legion officer should be, almost so typical that his manner might be thought a pose. But this Dane was as good as he seemed. He was one of that bevy of intrepid officers who even in those 'good old days' preferred the desert and the sand to the comforts and pleasures of garrison towns, despite the flattering conquests which their poise would have won them. For life to be really satisfying to Selchauhansen, however, one thing was missing – fighting.

During these early days of the twentieth century there was little prospect of his wishes being fulfilled; there was no sign of war on the horizon anywhere. 'Pacification' was now old history and Bou Amana's revolt between 1881 and 1885 had been forgotten. In fact Selchauhansen was getting a little bored.

Although penetration into the south of Oran Province continued, for thirty years nothing had happened in Algeria apart from the routine construction and occupation of posts. The Moroccan frontier was fluid and to delineate it France was using encroachment tactics. There was no question of genuine annexation; Britain and Germany were on the watch for any tendency to expansionism on the part of the colonial powers and the Paris government had no intention of risking complications. So, a French presence was established slowly and patiently; it would eventually be accepted, the soldiers hoped, by force of habit. From time to time some 'unruly' tribe from Morocco would attack a convoy. There would be a short engagement lasting a few hours at most and the attackers would vanish to the shelter of their frontiers, out of reach of reprisals.

Selchauhansen sighed. He could not make up his mind which was best: life in the posts where the peace was sometimes disturbed by some harassing enemy or that of the mounted companies which criss-crossed the desert escorting supply convoys to the posts in the south.

Life in the posts was monotonous. The bugler sounded reveille at dawn and the night sentries were relieved. About 9 a.m. a water fatigue went out to reconnoitre the area round the well. The day wore on with similar duties – colour-raising ceremony, guard inspection, lunch, siesta, barrack fatigues; there was always a wall to repair or a trench to be improved. At 5 p.m. came retreat, at 6 p.m. guard-changing and in the evening lights out.

As a precaution against boredom the Legion had invented a complicated ceremonial, respect for superiors being shown in every gesture, every movement. Every NCO must have been saluted twenty times a day – each time a man met him in the courtyard of the post and that was very often. If a legionary was called for, he would come up at the double, stand strictly to attention and start by announcing in a voice which made the walls reverberate: 'Legionary Hartwig [or Bronelli or Leroy], army number 82208, four years service, present, sir.'

But apart from this, what was there? Could officers and NCOs really be sure that they knew their men?

One of the Legion's favourite sayings was: 'Legionaries who are bored are on the verge of making BFs of themselves'.

'Making a BF of oneself' was both a vague and an explicit phrase; it covered any crazy notion which the desert – or the 'desert blues' – might induce in an empty head. It might mean desertion which sometimes, for no apparent reason, would decimate a platoon. It might mean suicide which would sometimes spread like an epidemic in an isolated post. It might mean a stupid and obstinate refusal to obey – a mere 'to hell with it' could land the culprit in the guard-room. It might be some spontaneous beat-up which would land some pig-headed legionary in the CO's orderly room. A cheeky little Belgian once said to Colonel de Corta, one of the great figures of the Legion in Africa: 'Colonel, you've got a great fat cigar. If you had in your mug what I've got in my pants, you wouldn't be able to say "papa".'

'Making a BF of oneself' was generally quickly dealt with by a well-aimed blow to the jaw. The Legion did not always settle its problems by reference to regulations.

Selchauhansen sighed again. 'There's nothing like a good scrap; the legionaries are beginning to get bored.'

For something to do it occurred to him to go round to the BMC, the travelling military brothel. This did more than satisfy a physical requirement; as an institution it was good for company funds since a proportion of the 'receipts' was paid into the company accounts and used to purchase additional amenities for the men.

Vauchez, the captain commanding the company, would often remark, 'So in the end everyone's happy.'

Vauchez was the precise opposite of Selchauhansen, a little lean man with a scraggy moustache. He was finicky and quick-tempered, drank hardly at all – whereas Selchauhansen's drinking bouts were a by-word – smoked little and slept even less. Vauchez and Selchauhansen complemented one another; the captain was insistent on strict observance of regulations; the lieutenant took actual pleasure in breaking them.

They were on marvellously good terms, though Vauchez would often grumble when irritated: 'You and your notions of heroism; you'll get all your men killed for the glory of it.'

This was not true, of course, but it was enough to allow Selchauhansen to pretend to be deeply hurt; he would reply:

'What with your mania for planning everything, nothing unexpected will ever happen to us. Even the Chaambas, and God knows they're incalculable enough, have had enough of trying to attack us.'

Vauchez shrugged his shoulders. He had no intention of repeating the experience of the French post of Taghit, on a spur of the Grand Erg, about sixty miles from the post at Bechar. Back in August Taghit had been surrounded by the Chaambas for five days. They fired at the legionaries from the top of the sand-dunes and prevented them reaching their source of water. For five whole days the yelling tribesmen, encouraged by the wails of the Berber women who had followed their men into the field, kept the post isolated, hoping for its surrender. There were 3000 of them. For five days Sergeant-Major Gabaig's legionaries held out behind their low 'toub' walls – dried mud mixed with straw. They had fought to the very last man.

Such battles may have excited hotheads like Selchauhansen but they left Vauchez cold. He was in command of a mounted company and that was all. Not that he was afraid; he could have remained in a garrison town like Brive-la-Gaillarde or Nîmes. He merely thought that an officer of the Foreign Legion, though he should have no fear of death, was, after all, of more use to his country if he remained alive. Also he was responsible for the lives of his men and he had no wish to lead them to death merely for the glory of his company. Vauchez knew that the Moroccans who crossed the frontier were no easy meat. If they attacked, their primary object was plunder and they would only attack if they were certain of winning, only against a force weaker than their own. They were brave but not foolhardy.

'They have not been brought up in the Camerone tradition,' he said but without a laugh.

Laughing was part of Selchauhansen. Vauchez heard the Danish giant's loud laughter shattering the calm of the evening. The BMC was close by, near the MO's dug-out.

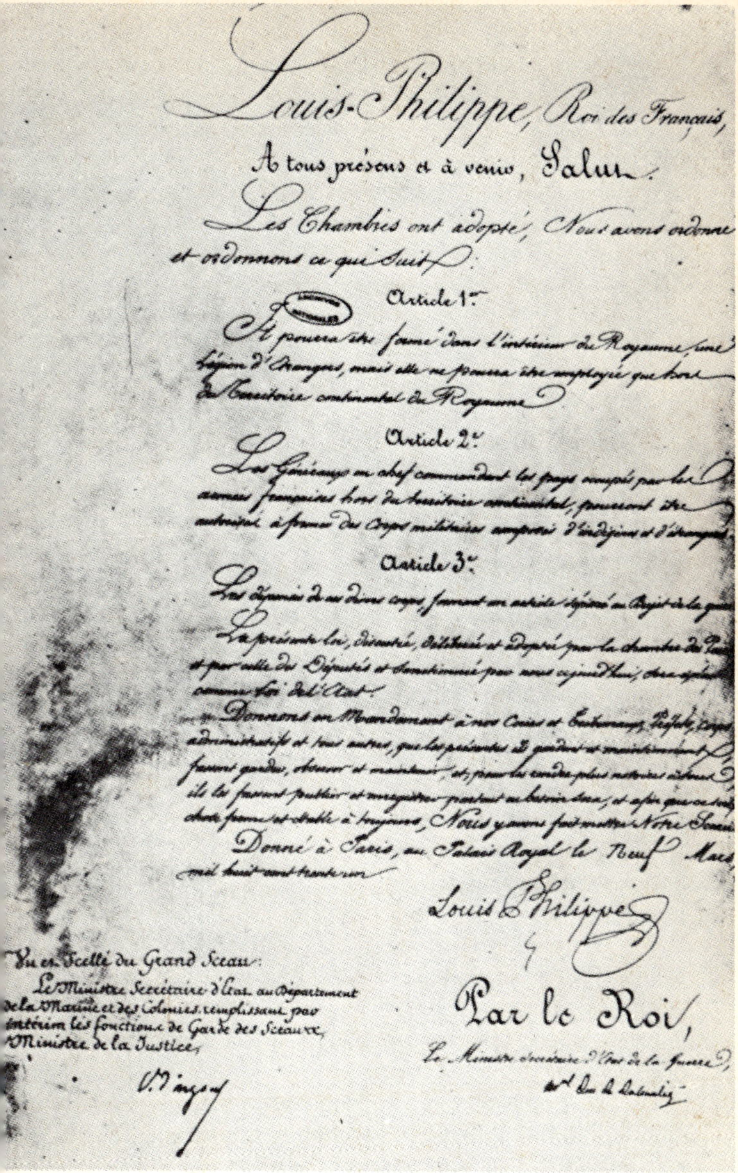
Louis-Philippe, Roi des Français,

A tous présens et à venir, Salut.

Les Chambres ont adopté, Nous avons ordonné et ordonnons ce qui suit:

Article 1er.

Il pourra être formé dans l'intérieur du Royaume une Légion d'Etrangers, mais elle ne pourra être employée que hors du territoire continental du Royaume.

Article 2e.

Les Généraux en chef commandant les pays occupés par les armées françaises hors du territoire continental, pourront être autorisés à former des Corps militaires composés d'indigènes et d'étrangers.

Article 3e.

Les dépenses de ces divers corps, feront un article séparé au Budget de la g

La présente loi, discutée, délibérée et adoptée par la chambre des Pa et par celle des Députés et sanctionnée par nous aujourd'hui, sera exécutée comme loi de l'Etat.

Donnons en Mandement à nos Cours et tribunaux, Préfets, Corps administratifs et tous autres, que les présentes ils gardent et maintiennent, fassent garder, observer et maintenir, et, pour les rendre plus notoires à tous, ils les fassent publier et enregistrer partout où besoin sera; et afin que ce soit chose ferme et stable à toujours, Nous y avons fait mettre Notre Sceau.

Donné à Paris, au Palais Royal le Neuf Mars mil huit cent trente un.

Louis Philippe

Vu et Scellé du Grand Sceau:

Le Ministre Secrétaire d'Etat au Département de la Marine et des Colonies, remplissant par intérim les fonctions de Garde des Sceaux, Ministre de la Justice,

Par le Roi,

Le Ministre Secrétaire d'Etat de la guerre,

Mal Duc de Dalmatie

. The Legion's birth certificate dated 9 March 1831: 'There may be set up within
e Kingdom a Foreign Legion . . .'

2. The final scene of Camerone: Maudet, Maine, Catteau, Wenzel, Constantin and Leonhart, having fired their last rounds, charge the encircling Mexicans with the bayonet.

3. Graves at El Mungar. The memorial to the battle in the background.

4. The Legion holding up the German advance in the Bannier suburb of Orléans on 9 November 1870.

In the Bou-Denib redoubt, destined to become famous.

Apotheosis of valour at the end of the bloodiest war France has ever known. olonel Rollet together with Sergeant-Major Mader and Corporals Arocas, Dieter nd Leva, all four wearing the Legion of Honour.

6. One of the Legion's trenches during World War I

7. The sortie from Bir Hakeim during the night of 10 June 1942.

8. A certain pomp and circumstance is good and the 'Foreign Royals' ensured that there was plenty of it: 1st Foreign Legion Cavalry drummer in 1925.

The BMC consisted of pretty girls recruited by an ample-bosomed Jewess from Oran – she had remained worthy of her nickname of 'Mama'. It was always on the move and came up as part of the supply convoys. The legionary lost to the world in some isolated post had not only to be well fed but also given his little pleasures. When the arrival of a convoy was announced, from the moment that they could see it the legionaries made ready to welcome it. In the twinkling of an eye, even in the most remote corners of the South, best uniforms came out, were ironed and brushed. Boots shone, moustaches were waxed and new *képis* taken from their boxes. It was a never-ending miracle, for most of the time water, water above all, was rationed for the 'prisoners of the desert'.

Vauchez went back to his tent and stretched out on his camp-bed, his hands behind his head. The operation for which he had been detailed was pure routine, except for one thing: the convoy of which he was in charge was the largest ever collected – 3000 loaded camels.

Obviously one company could not be responsible for all this. Overall command was therefore in the hands of the most senior officer, Bichemin, commanding the battalion.

The 3000 camels were divided into six groups or echelons, leaving every three hours. The wells had to be allowed time to fill after each group passed. Vauchez with his company, consisting of ninety-eight legionaries and a troop of mounted spahis, was responsible for the security of the second echelon.

The first part of the convoy was due to leave at midnight under command of Captain Bonnelet, a fellow-officer of the 1st Foreign Legion. Vauchez's turn would come at 3 a.m.

'May I come in, Captain?'

'Yes.'

Vauchez's orderly entered stooping, his tall silhouette filling the tent door.

'There is a message for you. Major Bichemin advises you to keep a good look-out. A *harka* [large group] of enemy Chaambas has been reported in the area of Zafrani.'

Vauchez made a wry face. Zafrani was precisely the direction in which he would soon be moving.

'Assemble the NCOs and go and find the lieutenant.'

Selchauhansen arrived. He was happy to hear that there was a possibility of a 'scrap' and winked: 'Let's hope Bonnelet leaves us some of them.'

Vauchez took no notice of this interruption. He issued his orders calmly; the company was to keep in tight formation.

Sergeant Perret-Dessus, the duty NCO, interposed. 'Captain, I must tell you that some of the camels have not kept to the timetable and have left as soon as they were loaded.'

Vauchez sighed. The Arabs guiding the *sokhars* [groups of camels] had disregarded the orders.

'This is going to give us a column over four thousand yards long. I'll be with the rearguard and make the tail-end camels gallop to catch up the main body.'

Selchauhansen took a philosophical view of the situation. A four thousand-yard column was vulnerable. There was no reason why Vauchez should have the ill-luck yet again to miss a proper battle.

'What about guides?'

'Here they are.' Perret-Dessus pushed forward two Arabs. They were in rags with a sort of faded red tarbush on their heads. They looked more like town beggars than denizens of the desert.

'The Zafrani track. *T'arfou?* Do you know it?'

The Arabs shook their heads. '*M'anarf* – I don't know it, I swear.' The second admitted that he must have used it once – 'But it was a long time ago, you know.'

'That's a pretty mess!' Vauchez sighed. 'And the night's as black as your hat. We'll be lucky if we don't get lost.'

'We'll do a Tom Thumb,' Selchauhansen chuckled. 'Catch up one group of camels after another and when we've got no more in front of us, we shall have got there.'

It was now 3 a.m. Vauchez rode up to the head of the column and gave the signal to move off. Rapidity and silence were the order of the day, for at night the least sound can be heard miles away. Perret-Dessus and de Montès, a Spaniard with the fine figure of the Andalusian mountaineer who was acting as quartermaster-sergeant, had accordingly checked that all mess-tins and weapons were securely fastened.

The company moved off without a sound. Far ahead the convoy was spread over miles of desert. The spahis scouted in front; the legionaries protected the flanks.

At 6 a.m. came the first half-hour halt. It was broad daylight when the company moved off again. Selchauhansen sent back a message to say that the formation was closing up.

'Good,' Vauchez replied. 'We'll try to concentrate the tail of the convoy at the next well.'

This was his main worry. With only 120 men he obviously could not even try to cover a column 4000 yards long. In the event of attack or any other contretemps the men would merely have been picked off individually without achieving anything. He had therefore decided to keep his company concentrated in platoons, ready to descend on any threatened point as necessary.

At 7.30 a.m. Selchauhansen reported that he had caught up 200 camels at the well and was holding them pending the arrival of the company. Vauchez duly arrived, ordered a halt and the filling of individual water-bottles and the mule-borne water containers. At 8 a.m. he gave the order to move. The company was now escorting 200 camels, the greater proportion of the second echelon.

The march continued under an increasingly hot sun. To right and left nothing was to be seen but a sea of pebbles, a vast reddish plain rising gently to some hills two to three thousand yards away. There was not a breath of air. They were not yet in the sandy desert, nevertheless this was the Sahara. Vauchez, however, was indifferent to the countryside. He stepped up the pace. In the distance he had seen a tiny black speck at the top of a rise – a few camels entering the El Mungar defile.

Somewhat closer were to be seen the white and red burnous of the leading spahis, spread out as scouts. They rode in groups of three, searching the little depressions scored out by running water – how long ago? – then concentrating and marking out the track. Then came the two hundred camels, swaying nonchalantly beneath burdens which made their backs look enormously out of proportion. Beside them trotted their guides, barefoot and in rags, urging them on with raucous shouts. The company brought up the rear, moving in two platoons – No. 3 under Selchauhansen and No. 4 under Danner, an Austrian sergeant, black-haired and swarthy, like some fearless placid beast. But appearances were deceptive; Danner was the best NCO in the company. Grouped around Vauchez was the little company headquarters – Sergeant-Major Tissier his second-in-command, de Montès the quarter-master corporal and his orderly.

The spahis had reached the El Mungar defile. It was a narrow basin, the sides consisting of gently sloping hills covered with thin tufts of alfalfa and scraggy bushes. At the entrance was a little gully, also covered in alfalfa, the bed of an old stream long since dry which curved round the basin in an arc.

Vauchez screwed up his eyes and surveyed the scene. He knew that there were only two possible points for an ambush on his route – here at El Mungar and a few miles further on at El Achrak, where a small detachment had been wiped out three years before. He saw the spahis moving up the basin, having crossed the gully. There was nothing to fear, therefore, for the gully was the trickiest point in the passage of the defile.

'Halt!' The command rang out echoing from the hills. This was the regulation halt. The legionaries piled arms and took off their packs. They were facing south. The mules were hobbled a little distance away, in rear of the company.

'Breakfast halt,' Perret-Dessus, the duty sergeant, announced.

This information was unnecessary. The legionaries knew the

military maxim: 'A legionary eats when he can and marches as far as he can; if he lies down it is to die.' There was no question of dying – at least not yet, so the legionaries ate.

While munching away Sergeant-Major Tissier, sitting near the captain, was keeping his eyes on the four flank-guard spahis who were scouting up the gully. Then the shots rang out. Two of the spahis fell from their horses, shot at point-blank range; the other two galloped back towards the Legion.

The men reacted at once. Arms were unpiled in a second and they closed on their NCOs, taking up a defensive position facing the firing. In front of the headquarters group Selchauhansen's No. 3 Platoon was in position on top of a little hillock. They were the nearest to the centre of the ambush and they formed a circle firing in all directions.

Selchauhansen stood up, calmly issuing his orders, directing his fire on to the Chaambas' positions which he could see only eighty yards away to his right.

'To the left, you set of fools,' he yelled. 'You're just spraying the countryside.' He crouched down, adjusted a rifleman's sights and stood up again.

He was hit three times simultaneously and fell. Two legionaries dropped their weapons and rushed to him.

'Leave me alone! Get on and fight! Fire, for God's sake, fire!'

'But, Lieutenant, you're wounded.'

'So what? That won't help you. Do your fighting first.'

The legionaries had not noticed that Selchauhansen had been hit in the head. As they were about to reply, they saw that he was dead.

'No. 4 Platoon, face north!' Vauchez had taken stock of the situation. The Chaambas were not only in position to the west; they were everywhere, forming a semi-circular line of fire following the line of the gully. The company was caught in a trap. It had to fight in all directions.

Danner and No. 4 Platoon turned about and opened fire.

'How many are they?' Tissier asked.

'About eight hundred,' Vauchez replied after a moment's thought. Tissier drew a long face. This was not going to be a piece of cake.

Vauchez summoned four legionaries and despatched them in pairs to warn the nearest posts – Taghit to the south, where Captain Bonnelet's first echelon should have arrived, and Major Bichemin who was coming up from the north with the rest of the convoy. It was now 9.40 a.m.

A hail of fire came from all directions, pinning the legionaries to the ground. It was eerie. Flying sand mixed with the smoke of the powder and turned into a red mist, clinging to the ground, blurring details, giving them a spectral appearance. The odds were six to one in favour of the Arabs. They were waiting for the return fire to slacken before rushing in to the assault. But the legionaries did not fall back.

The Chaambas manoeuvred, observing that between company headquarters and No. 3 Platoon the position was less strongly held; the latter was under direct fire and its losses were considerable. Some hundred burnous emerged from the gully and dashed forward yelling, followed by the wails of their women.

'Fix bayonets!' This came from Sergeant Danner of No. 4 Platoon. Realising that No. 3 was about to be overwhelmed, he moved out to their assistance. He crossed no-man's-land and fell upon the flank of the rebel attack. With him went Corporal Cachès' section which had been guarding the mules; it had a personal account to square with the Chaambas since they had just killed its lieutenant.

But against targets in the open the enemy fire from the gully was too hot and too accurate. Losses mounted and No. 4 Platoon's counter-attack was soon brought to a halt. The legionaries were pinned to the ground.

Vauchez, who was watching his men's advance from his command post, was hit – for the third time since the start of the attack. This time he was hit in the stomach and slid to his knees. Sergeant-Major Tissier, next to him, had also been hit. The two men sat down, leaning against one another.

There were now only two NCOs in the headquarters, Corporal de Montès and Sergeant Tisserant. Feeling themselves useless, they ran forward to join the fighting. They reached a position near No. 4 Platoon and took over from the legionaries who were charging forward. They eventually joined up with No. 3 Platoon which, with this unexpected reinforcement, was able to repulse the assault.

They had succeeded, but at what a price! All NCOs of both troops were either dead or wounded. Tisserant collected them around the survivors of No. 3 Platoon, near Selchauhansen's body, and himself took command of the remnants of No. 4 Platoon.

'You stay here,' he ordered de Montès. 'Hold on with No. 3 Platoon. I am going to manoeuvre to get back to the captain.'

De Montès organised his position, distributing the riflemen and caring for the wounded. Among the latter was the great Sergeant Danner, the giant of No. 4 Platoon. He had a bullet below the navel; he had pushed his whole forefinger into the hole to extract it. His face contorted with pain, he could only say, 'Montès, old friend, I believe I'm done for.'

Montès drew a long face. It was not only the wounded who were done for; it was the entire company.

The Chaambas now tried a new manoeuvre; they infiltrated between the two platoons in an effort to isolate them and reduce them in detail. Meanwhile Tisserant managed to get back to his starting point, a more secure position than that of No. 3 Platoon.

But the rebels made progress. At all costs they must be prevented from splitting the column. Tisserant despatched a runner to de Montès ordering him to withdraw. On the way back, however, de Montès was wounded by a bullet in the stomach. His place was taken by Corporal Detz.

It was now 10.30 a.m. The company had been fighting for over an hour.

They were now completely surrounded. The survivors of No. 4 Platoon were grouped around Captain Vauchez who was in terrible pain. The few survivors of No. 3 Platoon under Corporal Detz were further south, gathered round the body of Lieutenant Selchauhansen which his men were unwilling to leave (they knew the desecration which the Chaambas wreaked on the dead).

The rebels were massing for a frontal assault from the east. Vauchez saw them and called Tisserant: 'Form square. This is the moment to form a front.'

'And what about Detz?'

'Tell him to rejoin us. Holding his little knoll serves no good purpose. We might as well all die together.'

Tisserant called for a volunteer to carry the order. A spahi came forward – Peroni, the survivor of the morning's patrol. He leapt on a horse but it fell with a bullet in the head. 'Take mine,' Vauchez said. Peroni mounted and dashed off, pursued by Chaamba bullets. He passed on the order and returned at full gallop. Hardly had he reached the protection of the legionaries than his horse collapsed, hit in the heart.

Tisserant concentrated his diminishing force but he left on a little rocky spur three men, all it could hold, under Corporal Zoli, a withered little Italian, quick-witted and with a dry humour.

'Sacrificed for the sake of being sacrificed,' he grumbled, waving his rifle about. 'Might as well sell our skins as dearly as possible. I'll do a Camerone all by myself if necessary.'

Tisserant was back with Captain Vauchez, bringing some fifteen legionaries with him.

'What's the time?' Captain Vauchez drew back his sleeve. On his wrist was a large watch, held by a copper bracelet. 'Half past one.' The company had now held out for nearly four hours.

Detz had been unable to rejoin the main body. He was still on his little knoll with some thirty legionaries, all more or less severely wounded. In this hopeless situation he showed his true worth. He had six years' service and was generally held to be a good NCO, discreet and efficient but somewhat too self-effacing; he was strict with himself and serious as a priest. Very few had ever heard Detz shout; he spoke in a low voice so that people sometimes had to strain their ears to understand him. It was a trick of his which forced people to listen to him.

At this moment he was a changed man. The fighting and the difficulty of his situation transformed him. He was everywhere, running from one foxhole to another, shouting, hurling invective at the enemy, shaking his fist at them. The enemy replied, promising him all sorts of choice tortures. They concentrated their fire on him but, leaping agilely from stone to stone, Detz mocked them as they missed him. Each time he moved, the same scenario was enacted. Everyone joined in the game. There would be a Chaamba volley followed by a unanimous shout of 'Missed!' The legionaries began to laugh. Morale rose.

Further back, facing east, Tisserant was preparing for an enemy charge. The Chaambas had crawled down into a small depression and worked their way to within range. Now they tried to rush the position.

Tisserant fell with a bullet in the thigh but continued firing from the hip at ground level, mowing down Arabs as they came on. From his spur Corporal Zoli had seen the threat to No. 3 Platoon. He swung his three rifles round and ordered: 'Rapid fire!'

His men had the Arabs in their sights as if on the range. They were at just the right distance – eighty yards – the sort of shoot one dreams of. Every shot found its mark. The Chaambas were surprised by this come-back and withdrew in disorder, running for cover.

'What's the time?' 'Two o'clock,' Vauchez replied without looking at his watch. He had been counting the time and making a mental calculation. His messengers had left El Mungar four and a half hours ago. They should by now have warned the others who ought to come soon.

Nevertheless nothing happened and ammunition was running out – water too. The heat was suffocating, increased by reflection from the rocks.

Fighting went on until 4 p.m., assaults sometimes being repelled with the bayonet. The wounded were groaning, licking the blood from their wounds and sometimes wiping their burning faces with a handkerchief moistened with their own urine.

At 4.15 p.m. Sergeant-Major Tissier, who had stood up, was hit in the head. He collapsed, his brains spattering over Captain Vauchez's tunic. The latter almost fainted but kept a grip of himself, tight-lipped and with face lined by the struggle and by pain.

Half an hour later the situation still had not changed. Undoubtedly the Chaambas had realised that it was no good continuing to attack across open ground legionaries who, since the beginning of the battle, had had time to improve their positions. They had turned the bodies of their dead comrades into parapets and would not be moved for they would never abandon those comrades. So the Chaambas turned to harassing tactics and started shooting down the mules.

'They're off!' Standing on his position 300 yards to the south, Detz was shouting and his voice reached the legionaries of No. 3 Platoon who gave a sigh and looked up. Detz was right. In little groups the Chaambas were leaving their positions and moving away westwards. Morocco was not far off and they would be safe there.

'What's the time, Captain?' . . . 'Twenty past five.'

No shot had been fired for a full five minutes. There was complete silence in the little hollow, a silence all the more impressive by comparison to the previous infernal noise. The battle was over; it had lasted seven hours and forty minutes.

On Tisserant's orders a party moved out cautiously to the dead mules which had been carrying the squadron's water-containers, filled at the last well. But they had been holed by Arab fire. The party only brought back what remained in the bottom of the containers – a few pints which Tisserant doled out, with priority to the wounded.

'As far as we are concerned, we can wait,' he said apparently forgetting that he was wounded himself.

It was now 5.30. 'Cavalry to the south!' It was the *moghrazni*, Arab auxiliaries in French service, coming up from Taghit under Captain de Susbielle.

Two of the legionaries sent out by Vauchez in the morning had managed to get through the enemy lines and had arrived at the post in a state of exhaustion, having covered twenty-five miles of desert in less than four hours. De Susbielle had sounded the alarm at once and set off with his cavalry. It was the dust raised by their horses that had put the Chaambas to flight.

During the evening Captain Bonnelet's squadron, the one ahead of Captain Vauchez, arrived in its turn. It brought with it two medical officers so the wounded could at last receive attention.

No. 2 Company of the 2nd Foreign Legion had fought well. It had suffered thirty-three dead and forty wounded. Captain Vauchez was not to leave El Mungar; he died that evening having been assured that his men were in good shape and the wounded had been cared for. 'Selchauhansen should be happy,' he murmured. 'He succeeded in being a hero.'

What neither Vauchez nor Selchauhansen were ever to know was that on that very day a small boy in France decided to become a legionary. Some months back he had met the big Danish lieutenant in the house of some friends and had fallen for him.

'I'm going to be a Foreign Legion officer,' the little boy said to his father.

'A descendant of Louis-Philippe does not become an officer in a line regiment,' his father replied.

Nevertheless this descendant of the family of Orleans, a prince in his own right, kept his vow. His name was Aage, Prince of Denmark.

Great Leaders

Africa 1900–33

'SIR, THE binister.'

De Corta looked up frowning. Murraus, the police post sergeant, had a terrible German accent but even in spite of the accent he was clearly referring to a minister.

'What are you on about with your minister?'

'Well, Golonel, the telephone – it's just said a binister's coming to visit the post.'

'On a Sunday!'

De Corta was furious. Sunday was sacred as far as the legionaries were concerned – and the officers too. Anyway, what was a minister doing coming to Taroudant right in the south of Morocco?

'Couldn't the fellow stay in the Residency at Marrakech?'

De Corta commanded a battalion of the 4th Foreign Legion. It was a unit formed as an image of himself, workmanlike, tough, contemptuous of anyone who did not belong to it. The week was fully occupied with military matters, patrols, expeditions, ambushes, pacification; they were at it twenty hours a day, six days a week. This should be enough to prevent someone stealing their Sunday on the pretext of parading for some minister who was playing the tourist. Like many of his fellow-officers de Corta was no republican. His headquarters pennant, fleurs-de-lis on a blue ground, showed what his views were. Though he was a viscount and a true aristocrat, de Corta was a typical Legion officer who drank deep and cursed loudly.

Over the telephone he made his reasons crystal clear to the Residency. It was too late. The minister had already left. De Corta was furious: 'It's not even a minister,' he rasped. 'It's some Under-Secretary of State. Probably an ex-schoolteacher and a freemason.'

But de Corta was disciplined. At the entrance to the town the salute was given by a platoon in full dress – white tunic and black baggy trousers. In the barrack courtyard the duty company was drawn up in a square and the officers stood in line. In front was the regimental band. The ministerial car arrived and out of it stepped a little man in a morning coat looking black and ridiculous in the midst of the legionaries in their white. Throwing out his chest, with his cane under his arm, de Corta saluted and moved forward, towering over the little man.

The Marseillaise was played, followed by the 'General Salute'. Escorted by the commander, the minister inspected the troops, and then, stepping back a few paces, gave these German, Czech, Russian, Italian and Spanish legionaries a sententious speech dealing with secularism and education of the masses.

De Corta had had enough. He couldn't care less. Taking advantage of a short pause while the minister drew breath, he declaimed in a loud, deadly serious voice: 'We are all grateful to you, Mr. Minister, for this fine speech. If we did not already know our duty, it would have taught it us. In any case it has kept us amused.' And then without a moment's pause, turning to his battalion: 'On your marks the cross-country team. You will run for the minister's prize today. Now, my dear minister, how much are you giving to the winner? A thousand francs?'

'Well, that is . . . I hadn't thought . . .'

'Good. The minister is giving a thousand francs,' and then in a whisper: 'if that's awkward for you, my dear chap, I'll lend you the money.'

Never again did a minister take it into his head to visit the

Legion on a Sunday. To exonerate the Third Republic it should be recorded that the minister did repay Colonel de Corta his thousand francs.

De Corta had some fairly severe brushes with the authorities. In Indo-China he said to a certain general one day: 'If necessary my men are prepared to die for you but I refuse to allow them to play the clown for your amusement.' This particular general de Corta nicknamed 'the refuse of war' and ensured that the general was well aware of the epithet's author.

Following the example of Lyautey this contempt for anything which was not connected with the Legion became one of the characteristics of its officers; it became known as 'Papa's Legion'. When Colonel Maire, another of the great figures, heard that an officer of his term at the Military Academy had been promoted colonel after passing the Staff College and holding a number of staff appointments, he said to him icily: 'You would obviously be promoted; the only balls you've ever heard whistle by are ping-pong balls.'

For these officers the Legion became a way of life, not a theatrical performance.

The casual attitude these officers displayed to representatives of the civil authority stemmed primarily from the mistrust felt by the professional dealing with the realities of pacification and responsible for the lives of the men under his command, in face of the pusillanimity of the Republic which issued orders and counter-orders without thought for their effect.

The first to set an example was Lyautey.

In 1900 the French were still pushing forward into the south of Oran province. Tribes of marauding Moroccans, however, kept the area in a state of increasing insecurity in the certainty that they could do so with impunity. At this period the Kingdom of Fez was still a feudal country where only those tribal chiefs who had sworn allegiance to the king recognised any form of authority. There was no administrative structure and the king ruled only his capital. Outside its walls the country was divided into a number of autonomous provinces under tribal control. The maintenance of consulates or the despatch of representatives from Paris, London or Berlin achieved nothing. The hinterland belonged to unpredictable Berbers who lived off plunder and the slave trade.

Accordingly Lyautey, deliberately disregarding his orders, established himself at Bechar and extended his control into the interior of the country. Initially his purpose was merely to give the southern area of Oran province the security essential to enable French penetration to proceed. Starting from Bechar, he drew an imaginary line on the map running south-westwards. 'There's my security line,' he said. To hold it he called upon the Legion.

The Legion reacted with its customary speed. The security line encroached considerably on Moroccan territory, but never mind – the point was to be in position before the European capitals, London, Paris or Berlin, protested. Berguent, for instance, was constructed in sixty-four days. It was a typical Legion post, combining (relative) comfort with strategic requirements. It resembled a fortress about to be besieged. After Berguent came 'Colomb' Bechar and then, further south, transit bases for the mounted companies.

These mounted companies of the Legion bore no resemblance to the troops of *méharistes* or squadrons of spahis with their burnous flying in the wind. They were poorly equipped and looked more like the marauders they were chasing than parade-ground soldiers. Their main characteristic was their meagre allocation of animals – one mule per two legionaries. Commanding them was a young captain with bright eyes and a thunderous voice, hard-bitten and resilient as a Berber of the desert – Captain Rollet. He had chosen asceticism as his hallmark and he kept to it throughout his career.

'He was known as the "rope-soled captain",' says Colonel Maire, one of his ex-subordinates. 'He invariably wore a khaki linen jacket – he did so even on the Verdun front in 1917 –

no shirt, and on his feet Arab sandals held in place by two thin leather straps round the ankles. Sometimes he carried a sort of pink parasol.'

Rollet was just the man required for the mounted companies. He was always on the move, always sending out patrols along the frontier. 'Inside Morocco, of course,' he told his men. 'Outside we should have to wait for them to attack us.'

Very quickly he laid down regulations to suit the job – no halts on the march, no prolonged bivouacking. Each mule carried the kit of two legionaries, six days' rations, the animal's own feed, and water for the 'team'. Every hour came the order 'Change mounts' and the legionaries changed places.

For five years not a day passed without these mounted companies patrolling. They turned the security line into a genuine frontier which ultimately became more impenetrable than any complex and costly system of posts and fortifications. Nevertheless there were attacks and ambushes. Companies were annihilated; they vanished in the desert without trace, leaving their comrades nothing but the determination to avenge them. Paris knew nothing of these disasters or these battles.

In 1900 one company was attacked at Taghit, fifty miles from Colomb-Bechar. Five hours of fighting ended in the death of a sergeant-major and five legionaries. Three years later came El Mungar. Meanwhile Jonnart, the Governor-General of Algeria, had fallen victim to an ambush. The decision was taken to invade Morocco.

This was no full-scale landing with headlines in the newspapers. It was neither the blitzkrieg nor Suez. It was a stealthy penetration without fuss or publicity. The object was to impress the Moroccan government without offending the 'Allies', British and German. The latter reacted nevertheless and while the French were gradually establishing themselves in Morocco, William II sent two of his warships to Agadir. Paris vacillated but eventually stood firm, more because Lyautey had decided to remain, come what might, than from any real desire to conquer the country.

It was not a bloodless conquest. At Menhaba Colonel Piéron's column of 3000 men was badly mauled by an army of 30,000 Berbers. At Bou Denib the battle for the recapture of the town which the legionaries had seized was so savage that Colonel Alix's relieving column, which arrived after a thirty-six-day siege, could barely enter for the stench of the corpses both outside and inside.

But gradually the French gained ground. The Sultan Abd el Aziz was overthrown by his brother Moulay Hafid. He was besieged in Fez and appealed to the French. Lyautey rushed to his assistance and freed him. The French protectorate over Morocco had been established.

In 1914 came World War I. The Foreign Legion departed to the French front, leaving in Morocco only its German and Austrian nationals. Theirs was the responsibility for a country still in a state of ferment, with revolt brewing.

On the outbreak of war Paris sent orders to Lyautey: 'Withdraw to a security line; the question of reoccupation will be decided later.' Lyautey accepted this edict but he did not obey it. Not only was there no withdrawal but he pursued his pacification policy indefatigably. He knew that he had a free hand. No one would come from Paris to check whether orders issued would be carried out.

It was a difficult, thankless and unrecognised task; all France was looking eastward and cared little about Morocco. Lyautey held on for four years with two depleted regiments of the Foreign Legion assisted by two under-strength mounted companies.

When the war ended the Foreign Legion resumed its place in Morocco. The country was divided between two regiments – the 3rd Foreign Legion which inherited the battle honours of the Legion's wartime regiment, the most highly decorated regiment of the Great War, and the 4th Foreign Legion. In fact these units soon lost their numbers and were known by the names of their commanders – the de Corta battalion, the Nicholas battalion, the Tcharner battalion and so forth. Each battalion became heir to the halo of legend surrounding its commander.

De Corta, for instance, was a captain on the outbreak of war. He left Algeria in September 1914. He was wounded and taken prisoner almost at once and then for four years this aristocrat, whose trade was war, gnawed his fingernails at his inability to fight. He escaped, was recaptured and sent to a punishment camp.

'An officer never stoops,' de Corta said. He applied this maxim even to the most mundane of actions and contrived to impose the rule even on his gaolers. Drawing himself up to his full 5 ft. 9 in., he announced that 'taking one's boots off involves stooping'. His gaolers were so impressed that one of them removed his boots each evening. When the war ended one of these gaolers followed 'his' major into the Foreign Legion.

It was a strange unit, this Foreign Legion of 1919. It united under the same cap-badge men who had been fighting each other on every front for four years with the ferocity of irreconcilable enemies. Here they all were, ready to make war again – Ukrainians from Wrangel's or Denikin's White Russian armies, Bulgarians, Austrians, Germans, all with no other prospect than joining the ranks of the unemployed at the end of the war. There were Italians too, Belgians, Americans, British and even some Frenchmen, all of whom simply wished to go on fighting. They came together and fraternised, ex-enemies who now became brothers in the insecurity of a little post in the Atlas mountains.

During these years when the French grip over Morocco was being reasserted, the regiments of the Legion held isolated posts for the greater part of the year and only engaged in major pacification operations during a short period in the spring. The only object of these operations, moreover, was to establish additional posts in order to consolidate the French presence.

The Foreign Legion post in Morocco was by no means the palace pictured in films of the period which featured sand, palm-trees and Berber dancing girls with sly smiles and come-hither eyes. The 'legionary smelling of hot sand' was no more than a pleasing story. The reality was more prosaic and more monotonous too. Pleasures were few, the loneliness total and 'desert blues' ever present.

'The blues' was one of the Legion's most serious diseases at this period. It was responsible for the most incalulable happenings ranging from an ordinary beat-up, possibly involving some crime, to the inexplicable suicide.

In 1923, for instance, a young 2nd-lieutenant and a few legionaries were sent to hold an isolated post on the edge of the 'chleuh' country looking out on to the vast desert area to the south. During their first month all went well. The 2nd-lieutenant was an enthusiast and his men were disciplined and bent on their work. Then one morning a laconic telegram arrived in Marrakech: 'Legionary Greber dead. Stop. Hanged on his mule-chain.' There may be all sorts of reasons for a legionary to commit suicide – personal, emotional or simply 'the blues'. One could never tell in the Legion.

Headquarters ordered that an enquiry be instituted pending the arrival of the next supply convoy when the matter would be cleared up.

On the spot the 2nd-lieutenant arranged a little cemetery a few yards from the post. A regulation grave occupied a space beneath the shadow of the whitewashed walls under the burning sun. Throughout the day the sentry on duty could see the little white wooden cross. Shortly afterwards came the second suicide. 'Legionary Krosta deceased. Stop. Hanged on his mule-chain.'

At last Marrakech took action. De Corta realised the danger of a suicide epidemic spreading. The platoon might be decimated by 'the blues'. He set out at the head of a column, heading for the post at top speed. When he arrived he found a garrison which had gone to pieces and a 2nd-lieutenant who had shot himself through the mouth.

So, to avert 'the blues', the Legion was happy that its men should drink. Dust, thirst and 'the blues' were the main reasons why the Legion became the archetype of a unit consisting of inveterate drunkards, of morose or picturesque alcoholics, sometimes dangerous, frequently incurable. There is a well-known episode of a legionary who suddenly went mad as a result of 'the blues' or drink and during a parade snatched up his rifle, aimed and fired at his platoon commander. Fortunately he missed. 'Fifteen days' detention,' the officer noted calmly. 'Reason: missed his target at less than fifteen yards.' Faced with 'the blues', the Legion was invariably indulgent.

'Next to "the blues",' says Colonel Maire, 'enemy No. 2 in the posts were rats. They were enormous. None knew where they came from but they ate our reserves and might well produce typhus epidemics.'

To deal with these animals the Supply Service, with its mind primarily on its provisions, one day suggested a bonus of ten centimes for every rat caught. Each post was given an allocation of official rat-traps. The legionaries found the bonus highly attractive but they took little stock of the traps. Being of an inventive turn of mind, they soon found other methods of ensuring that the bonus of ten centimes per rat would bring in a tidy sum. Maire records: '278 rats were caught in a single night.' But the Supply Service proved a bad loser and refused to pay.

The legionaries shrugged their shoulders. They had already found something else. Tourists from Colomb-Bechar had begun to arrive to visit this famous desert, a place hitherto unknown but about which all sorts of tales were circulating, the latest being that it held the secret of Atlantis. They returned delighted, bringing with them, primarily in the case of the

Americans, stories of a species of prehistoric rat which soon became the source of profound puzzlement to the scientific world. 'At Colomb-Bechar,' the story ran, 'there still exists a race of rats to be found nowhere else in the world. On their noses they have a sort of protuberance like a rhinoceros's horn. Sometimes in fact the outline of a second protuberance can be seen in front of the first; its texture is strange, soft and sharp.'

The legionaries were commissioned to catch some – oddly enough these prehistoric rats would come near no one else. The price of a 'horned rat' rose sharply. Rat-catching almost turned into an industry and for a time the scientific societies were afraid that this intensive hunt would put an end to this rare species. Nevertheless horned rats continued to be found, but nowhere other than in Colomb-Bechar.

Six months went by. Reports and messages flashed back and forth. It was noted that, if a male horned rat was mated with a female horned rat, the result was an ordinary rat. Enquiries and searches went on.

The mystery was eventually unravelled by the Legion's garrison commander. Among his men was an ex-biologist who had a taste both for practical joking and money-making; he was grafting on to each animal's nose part of the tail of another rat. The graft was so well done that for months none of the experts had realised that they were being fooled.

Civilians were not the only targets for practical joking. In these posts the standard of comfort was extremely moderate; they resembled Arab shacks rather than oriental palaces; yet everything had to be accounted for, rationed, prescribed by regulations. The army was poor, as was the Legion, and the administrators were finicky. The administrative officer with his mania for returns and reports soon became one of the legionaries' enemies.

If the Legion was constructing a post, for instance, six months before work started a complete list of the material to be used had to be submitted, a plan of the locality drawn up and the definitive total of the cost of the work calculated. Accordingly the Legion started building first and asked for authorisation afterwards.

'It is no good fighting the bureaucrats,' de Corta explained. 'They have to be neutralised. For this purpose I make use of one or two clerks, ex-solicitors who are experts at drawing up reports. They account for and list everything conceivable. If the administrators protest that something has been done without their consent, I send them by return of post a forty-page report, prepared in advance, on the theft of a bicycle-bell. I enclose some equally voluminous report of a fire in some tent. By the time the administration has worked its way through all that, three months have gone by and my work is finished.'

As part of their campaign the legionaries would tell the story of the administrative officer who went on a tour of southern Morocco and was captured by rebel Chaambas together with his escort and a female camp-follower of twenty years' service. He was freed a few days later by a column of the Legion and questioned to find out how the ambush had taken place. The woman gave a graphic description of marches under the burning sun, lack of food, insults and threats. Finally they had arrived at the Arab camp. 'There,' she went on, 'we were stripped naked and tied to posts fixed one beside the other. The Sudanese then beat us and spat in our faces. Then we were forced to kneel down with our heads on the ground.' The sequel can be imagined. The woman added maternally: 'As far as I was concerned it hardly mattered. But this poor gentleman from the Supply Service – he wasn't used to that sort of thing. I was really sorry for him.'

The Supply Service sometimes showed that it possessed a sense of humour, however. At this period no scrap of material was allowed to be lost, at least not on paper; everything had to be accounted for logically and the loss of a tent-peg called for a report in sextuplicate.

The Legion had better things to do and took advantage of

natural catastrophes to 'lose' anything which had in some way disappeared. In the south, however, natural catastrophes such as avalanches of snow or floods are rare. But there were sandstorms and these the authorities would accept as a reason for losing things provided the report was accompanied by a local meteorological bulletin.

One day a report was received from the 1st Foreign Legion Cavalry Regiment, then in south Tunisia, listing losses incurred during a particularly violent sirocco. Among the tents and other objects which had 'taken to the air' appeared, to everyone's astonishment – an anvil! It belonged to the farrier-sergeant and had taken wing no one knew whither.

The Supply Service accepted the entire list without a murmur but against the word 'anvil' noted: 'Agreed in view of the abnormality of the occurrence.'

Admittedly these are mere anecdotes but they stand out in comparison to the daily routine to which life in the posts was confined. In these an assortment of men were condemned to live constricted to a small enclosure surrounded by mud walls, doing the same things over and over again day after day.

For the first few days everyone was on his toes: rebel marksmen were close by and woe to the sentry who lit a cigarette during the night; a bullet in the head was his reward for being careless. Then gradually, with nothing to be heard or seen, people relaxed and let things slide – until caught out next time.

During the days when nothing happened it was difficult to maintain order, to insist on customs which seemed meaningless, a mere tradition or liturgy rather than a necessary duty. An old Legion saying ran: 'If the legionary gets bored, he makes a BF of himself'. This was even more true of those posts where there was frequently nothing to do other than mount guard and polish the same gun barrel for the twentieth time until one was sick of it.

Desertions were not unknown. In Morocco, with the sea close at hand, men dreamed of escaping. The proximity of Spanish Morocco was an invitation to adventure. More serious were the occasional activities of enemy agents, particularly Germans during the years before World War I, whose purpose was to instigate mutiny.

In his memoirs Colonel Maire tells of one instance which nearly cost him his life during the winter of 1914–15, and of his way of dealing with it. 'I had sat up late with a friend that evening and did not get back to my room until after midnight. I noticed a bullet-hole in the window-pane. Next morning I found the print of a regulation boot at the spot from which the shot must have been fired. An expert Arab tracker made a careful study of the layout of the nails, noting which were there and which were missing. Eventually he told me: "I know who that boot belongs to; it's a certain von M . . .".'

Maire had the man in question watched and eventually learned that he had threatened to kill him – 'The captain must be done away with.'

Maire said nothing but one evening, when von M . . . was on guard, he had undoubtedly been drinking. 'Into the cells,' Maire ordered. He knew that it was no use bringing the German before a court martial since there was no proof, so he took matters into his own hands.

'The truth,' he writes, 'can be described in two words or gasps – the two final convulsions of that bastard von M . . . who died in his cell that night. Perhaps he had a stroke or perhaps it was paralysis.'

The funeral took place the same day and the entire company paraded to salute the coffin, whereupon Maire came forward and said: 'We do not salute blackguards.' Then, taking care to have his words translated, he explained: 'I have told you to take care of your health. In a climate like this, where the cold at night can be intense, rashness gets its reward.'

The 'mutineers' did not move a muscle. They knew what he meant.

'And then,' Maire went on, 'the ground is too hard for digging when it is frozen – even to put a captain under the sod.'

The company dispersed and never gave its commander another moment's anxiety.

Legionaries understand methods like that. Moreover they accept them because they feel the need both to fear and to admire. They are tough and they are happy to obey men tougher than themselves. Once they have made up their minds about a commander, they worship him. They look upon him as a demi-god. One such demi-god was Aage of Denmark.

'Sir, I am the son of the Prefect of Constantine and I should like a little more peace and quiet in this place.'

This was said in the Hotel Aletti, the biggest hotel in Algiers, to a major in the Foreign Legion who was making a great deal of noise; he was disporting himself in his usual way – royally. The major got up; facing him was a small, pallid, well-groomed young man, very much the socialite and with the supercilious air of the upstart who thought that his father's position placed him a cut above everyone else.

'Sir,' the officer replied, 'I am Aage, Prince of Denmark and grandson of Louis-Philippe, at present commanding a battalion of the Foreign Legion, and I make no bones about it.' Then he added tranquilly: 'Which way do you prefer to go out? Through the door or the window? I would add that, in the latter case, it would give me great pleasure to assist you.'

The Legion has a fund of stories about Aage of Denmark, some of them highly picturesque. When commanding a post as a young officer, he was confronted with the most surprising of situations. In one night two of his men deserted after killing a third. The fourth, a Russian, went mad.

One day, on return from a little 'relaxtion' in Marrakech, Aage of Denmark received a message: 'Commander-in-Chief visiting you this morning. Request you organise parade.'

Aage protested. 'Last night was pay night. All my men drunk. Impossible to guarantee organisation of parade.'

The order was not modified, whereupon the Major/Prince addressed his men: 'Look at me, you collection of clots. I am as tight as you are but it doesn't show. If I want, I can goose-step past.'

He proceeded to demonstrate. Thunderstruck the legionaries watched their commander as, erect and faultless, he showed them how a legionary should behave when parading in front of a general.

Never has a general been so beautifully received as the commander-in-chief was that day. Turn-out was impeccable and the march-past perfect. The general never knew that every one of the 400 legionaries who presented arms to him was drunk.

Aage of Denmark had chosen his profession years before – on the day when a young officer of the Foreign Legion, tall, blond and magnificent, had appeared in the Paris drawing-room where his mother was receiving. It was Lieutenant Selchauhansen, the hero of El Mungar.

In 1922, when the bank holding his family fortune crashed, he went to André Maginot, the French Minister of War, to say: 'I would like to join the Legion.'

'Your Highness, it would be an honour for the Legion.'

Very soon this Prince of the Blood became one of the Legion's most popular officers. He was both an aristocrat and a condottiere; he impressed everyone by his generosity – he would give caviar parties, – an extremely rare occurrence in a unit where officers lived on their pay – and by his forthright manner.

He was both reserved and relaxed. He captivated people by his brainwaves, enchanted them with his outrageous ties and astounded them by his courage. He was just as likely to take over from the drummer of an orchestra in Casablanca, in full uniform and *képi* askew, as to fight one of the legionaries of his company with his fists. In the latter case, to ensure that the bout was even, he took off his uniform and officer's stripes.

He was as devoted to his men as they were to him and his reminiscences often showed traces of emotion. He was a welcome visitor to all the courts of Europe, where he would tell of his battles and his men. One day he declaimed to King Leopold of the Belgians: 'And when the legionaries have drawn their pay on return from operations, they have a truly royal beat-up.'

Although a battalion commander, he could not take part in the war in Europe since in 1939 Denmark was considered non-belligerent. With a sinking heart Aage of Denmark saw the men he had loved so much leave for a war that was not his. He died in his bed of a heart attack. His last wish was to be buried at Sidi-bel-Abbès in the Legion's military cemetery.

De Corta, Maire, Aage of Denmark: they were all figures who left upon the Legion an imprint now outdated, an image of men of the desert, rough, tough, outspoken, formidable drinkers, impossible characters. They were 'sheiks' who demanded everything of their men, but never anything which they could not do themselves. Even at the risk of a serious breach of discipline they would defend their men.

In 1919 there was a captain in the 3rd Foreign Legion named Salomon. He was an old hand; he had fought in Tonkin as a legionary and then as an NCO; throughout the Great War he had been with the Foreign Legion Mobile Regiment [RMLE – Regiment de Marche de la Légion Étrangère] where his brilliant actions had earned him the epaulettes of an officer.

One evening Salomon's company was bivouacking with the rest of the regiment in preparation for an attack on Djebel Baddou due to take place next morning. As was the Legion's habit, as soon as packs were on the ground and rifles piled, each man went about his business; there was no need for definite orders. For an officer who did not know the form the impression was of total confusion but it was in fact the result of long practice; one squad was responsible for erecting the bivouacs (they had to be buttoned together); other groups dealt with the fires, shelters, sentry posts and latrines. In short everyone was busy with the chores necessary for living in the open.

Preceded by Major Nicholas, the battalion commander, a general arrived. Salomon came forward, looking very smart but with his chin stuck out. He was vexed; a general does not normally come to inspect while bivouacs are being set up.

'Fall in! Attention!' he shouted, however, in accordance with the rules.

The legionaries stood stock still as if rooted to the spot, their arms still full of tools, bundles of wood and equipment.

'Stand at ease,' the general said. 'Carry on.'

He looked round disapprovingly, clearly showing his dislike for all this activity, of which he grasped neither the scope nor the purpose.

'What's the meaning of all this flurry?' he asked Salomon drily.

The captain went pale. 'General,' he threw back, 'there was no flurry here until you arrived.'

The general took offence but Salomon refused to apologise. Finally Major Nicholas intervened, drew the general aside and explained the situation to him, stressing Captain Salomon's record of service. The general finally calmed down.

But Salomon was a character. He refused to reappear before the general who wished, if not to excuse himself, at least to say that he was ready to 'let bygones be bygones'. Moreover a few days later he said to his battalion commander with a surly air: 'If the general had gone on, I would have drawn my revolver and shot him. Then I'd have blown out my own brains.'

Colonel Maire, who tells the story, adds: 'I am convinced that he would have done it too.'

With the end of World War I the Foreign Legion returned to its duties in Africa. Its time was divided between Morocco, where pacification still proceeded, sometimes consisting of mere routine but frequently involving operations of war such as the reduction of Taza and the Rif rebellion, and Algeria, where recruits continued to arrive at Sidi-bel-Abbès.

There, after the return of the RMLE from the war, in which it had covered itself with glory, 'Father' Rollet, who had been promoted colonel in the field, ruled over the 'Mother Convent'. Before his arrival it had been merely a barracks, a recruit training centre and a transit camp for men on release. With the arrival of Colonel Rollet Sidi-bel-Abbès found its true vocation – that of a military and lay cloister where bodies and minds were moulded. Both officers and men were different once they had passed through Sidi-bel-Abbès. This transformation was due entirely to Colonel Rollet.

Physically the colonel had hardly changed since the time when he had ranged over southern Oran at the head of his mounted companies. He was a small spare man, his face scored with scars and with a piercing eye. He was a man of few words but he knew how to sum a man up at a glance and at the same time turn him into an attentive and devoted slave.

Every man's life as a legionary began with a traditional ceremony – the parade in front of the colonel. Doty, the American who can hardly be accused of having much love for the Foreign Legion, tells the story of this encounter:

> Colonel Rollet! I have heard much of him already and now know still more. Grand Commander of the Legion of Honour, Colonel of the 1st Foreign Legion, he is truly the 'Grand Old Man' of the Legion in which he has com-

manded (first as a young lieutenant) for nearly thirty years. During the World War he commanded a battalion of the 1st RMLE, a body thrown into every one of that war's worst furnaces and close to complete annihilation over and over again. Covered with scars and decorations but uniformed so simply that it is a standing joke with his men that he draws the same outfit as they do, he is never without the Legion's lanyard across his left shoulder.

He now came down the line, a little broad-shouldered man with a spade beard and fiercely turned-up moustache streaked with grey, and a nose like an eagle's beak. Before each man he stopped, sweeping him with a glance that took in every detail of his person and equipment, then asked him his name and nationality and went on.

And finally he was before me and I saw his eyes, his legendary terrible blue eyes. They had passed rapidly over me, he had asked me my name and my country and I had answered; now they were searching my own with that steady gaze of his like that of an old falcon. I looked him right back in the eye. That is what he wanted. '*Bien*,' he said shortly and passed on to the next man. Each he searched thus and to each steady look back, '*Bien*' he would say and pass on.

In that brief moment Legionary Doty had been captivated. He was not to be the only one.

The year 1931 was the centenary of the formation of the Foreign Legion. Rollet resolved that there should be more than a mere commemoration ceremony. He decided to endow the Legion's cradle with a war memorial which would be both an abiding symbol of fidelity to old comrades and a gesture of piety towards those foreigners who had fallen under the banner of the Legion. Rollet commissioned Major Maire, his old friend and second-in-command, to construct it.

Though the plans were drawn up by architects, actual construction was in the hands of the legionaries.

Maire reopened an old onyx mine some forty miles from Sidi-bel-Abbès. While it was being reactivated, a gang constructed in record time a motorable road between it and the barracks. Work began.

The foundation stone was laid on 8 October 1930 and the memorial completed in March 1931.

'We worked like Egyptians,' the legionaries said. They were no less proud of it for that. They had turned themselves into quarrymen, stone-cutters, masons, decorators, sculptors and carriers.

'The bill comes to 600,000 francs,' the works director told Colonel Rollet.

Rollet shook his head. 'And the contributions?'

'Not a sou.'

So Rollet took the inevitable decision. In his eyes this memorial was no more than was due to the memory of the 'foreigners who had become Frenchmen by shedding their blood', but since no one was willing to finance it, it should be an offering from the living legionaries to their dead comrades. For four years all legionaries, NCOs and officers, whether in Morocco, Tonkin, Madagascar or Algeria, gave up one day's pay per year.

'We will not be beholden to anybody,' Rollet concluded. 'The Legion does not go begging.' He had a list of the contributors – more than 30,000 of them – built into the masonry at the base of the monument.

In 1936 Rollet left Sidi-bel-Abbès having reached the age limit for his rank (a few years earlier he had been promoted general and appointed Inspector of the Foreign Legion). But the Legion did not lose him altogether. On his retirement he organised Old Comrades associations and to the very end continued in his role as 'father' of the Legion.

Then in 1939 came the war. Rollet wanted to leave his general's stars behind and revert for a few months to the rank of colonel commanding the 11th Foreign Legion Battalion, formed in France to fight against Germany. But he was not destined to take part in the 1940 defeat. He was considered too old to command and forced to relinquish his uniform for the last time.

1941. The Germans were in Paris. For two years they had been parading the streets of the capital as victors, making the pavements ring with the stamp of their jackboots.

For all these two years a certain window had remained firmly closed. Behind the drawn shutters, immured in his silence, was an old man who refused to go out. Never and nowhere had he been defeated and he could not bear to meet soldiers other than his own. At the age of sixty-seven General Rollet had not capitulated.

But he was sick and the name of his sickness was shame. He had been wounded too often and suffered hardships in innumerable campaigns. Now his health was fragile but he refused to go to bed.

'A legionary only lies down to die,' he kept saying in his dry tone. 'I am waiting for victory, the victory which France will inevitably snatch back one day and soon.'

Victory! As if anyone could be thinking of victory in 1941!

Nevertheless General Rollet continued to hope. All his life he had refused to give up; all his life he had pitted his determination against desertion, cowardice and weakness. He knew that he was right, that he would be right. But on 16 April 1941 he died without having heard that in Egypt the Foreign Legion was back in the battle.

At his funeral a few days later an erect, motionless, withdrawn figure stood alone in the centre of a space left by the mourners to underline the gulf which separated them. It was a German general in full uniform. General von Stülpnagel, Military Governor of Paris, had requested the honour of being present.

He was a well-known anti-Nazi and was hanged two years later for his part in the anti-Hitler plot. On this day he attended to render a soldier's homage to a French general who had been so good a commander of Germans. The legend of Rollet had penetrated even beyond the Rhine.

Auberive

April 1917

JULY 1916. Held in front of Verdun, the Germans had initiated a battle of attrition which proved a serious drain on Allied resources. The High Command felt impelled to relieve the enemy pressure and, even more important, straighten out the front line which formed a salient pointing menacingly at Paris.

The decision for the Battle of the Somme had been made. Twenty-six British divisions under General French and fourteen French divisions under Foch were to attack on a front of forty-five miles – general direction: north-east, in other words towards Péronne.

The date planned for the offensive was 1 July. Artillery preparation started five days before 'D-Day'; it was intended to crush the German front line under a storm of fire and steel, destroying the obstacles placed in the way of the Allied drive.

The RMLE,[1] the Foreign Legion Mobile Regiment, had only a waiting role in this offensive. The division to which it

1. This regiment [Régiment de Marche de la Légion Étrangère] had been formed on 11 November 1915 from four foreign regiments which had already fought on all fronts from the Argonne to the Marne, from the outbreak of war in August 1914 to 1915; they were the 2nd Battalion of the 2nd Foreign Legion and the 3rd Battalion of the 1st (formed in Paris), the 4th Battalion (or 'Garibaldi Legion') and 2nd Battalion of the 1st Foreign Legion, the latter formed of volunteers recruited in Lyon, Avignon and Bayonne. In three years the RMLE became the most highly decorated regiment in the French army. When the war ended its colours were handed over to the 3rd Foreign Legion Infantry (3 REI – 3 Régiment Étranger d'Infanterie).

belonged, the Moroccan Division commanded by General Codet, a Zouave, was in reserve some miles behind the front.

The August 1916 issue of *Illustration* gives a fairly accurate picture of the terrain – a vast devastated plain, pockmarked with debris and shell-holes and scored with trenches full of dead men, scraps of metal and general rubbish. Under the July sun it gave the impression of a frightful forty-five-mile-long cesspool. What the photographs could not reproduce was the smell but it can easily be imagined – an all-pervading odour of putrefaction. War had ceased to be a joyous affair.

For five days the artillery pounded the German lines, disrupting trenches, obliterating hills, filling in shell-holes as others were made. Then on the morning of 1 July the forty divisions charged forward. The frontage of the offensive had been reduced to twenty-five miles to give it greater weight.

The imagination boggles at the thought of twenty-five miles of men in uniform rushing forward with fixed bayonets through the chatter of machine guns, explosions of shells and bursting of grenades. Shouts interspersed with bugle calls were to be heard, for the bugle still sounded the charge. Twenty-five miles of soldiers charging ahead!

In front of the Legion the Colonial Infantry bore down upon its objective, a little village named Assevillers, or rather what remained of it, for there was nothing but a few remnants of wall, a sewer of a street and some skeleton trees. At 9.30 a.m. the colonials reported: 'Objective captured.'

This was quite a feat. Despite the artillery bombardment the Germans had resisted savagely; the very intensity of the five-day artillery preparation had warned them; they had evacuated their front line and reoccupied it as soon as the guns had lifted. So, where the forty Allied divisions had hoped to penetrate without firing a shot, they met an unpleasant surprise and were stopped almost everywhere.

At last the Foreign Legion received the expected order. It was to take over from the Colonial Infantry and the 39th

Reserve Infantry (Bretons). On 4 July the RMLE attacked from Assevillers. Its objective was another village which the battle had erased from the scene but which was called on the map Belloy-en-Santerre.

As it happened, some months earlier the Germans had selected this village, which was exceptionally well placed, as the linchpin of their defence. It looked down across an open slope about a thousand yards wide and had been well fortified with pill-boxes dug into the ground and flanking machine guns. Attacking Belloy-en-Santerre in broad daylight was tantamount to throwing oneself deliberately into the enemy cross-fire.

'That's why they've chosen us,' the NCOs said. They did not specify whether they were proud of the fact or whether they bore a grudge against General Berdoulat who had detailed them for this suicidal mission.

Nevertheless the legionaries of the RMLE were by no means downcast. Silently, during the night, they moved into their assembly area, ready to go. On the regiment's right were two companies, Nos. 9 and 11.

'Forward.' The bugle sounded 'Le Boudin' and then the quicker notes of the charge. The companies moved forward together. The ground was uneven and covered with half-burnt brambles; it was difficult to move at the double. So, in open order, the legionaries advanced on Belloy-en-Santerre. The enemy machine guns had not opened up – not yet; there were still 800 yards to go.

Then, when the companies were 300 yards from their objective, the Germans opened fire. A murderous enfilade fire struck No. 11 Company and many men were wounded at once. The company lost all its officers, mown down in the space of a few yards, and then its NCOs as they came forward to take charge.

The company was pinned to the ground among the weeds of an uncultivated field; the men crawled along, calling to each other, trying to re-form; the wounded were crying out, the stretcher-bearers, overwhelmed, were doing their best. Suddenly from the village came the sound of the bugle playing the famous 'Boudin'.

The story is told by Captain Tscharner, a Swiss baron who later covered himself with glory in Morocco. At once, he says, the legionaries leapt to their feet and, shouting '*Vive la France*', charged forward again. In a few moments the southern corner of the village had been captured.

Nevertheless, despite its forty divisions, the offensive made only halting progress. It continued for a further four months, but at what a price! On 5 November, when the Allies finally decided to call it off, the Somme front had been penetrated to a depth of only six miles at most.

Among the casualties of this attack was Alan Seeger, the American poet (army number 19522); the evening before the attack he had written:

> I have a rendezvous with Death
> At some disputed barricade,
> When Spring comes back with rustling shade
> And apple-blossoms fill the air –
> I have a rendezvous with Death . . .

In a few hours at Belloy-en-Santerre the Legion lost 25 officers and 844 men – a third of its strength.

April 1917. Having been halted on the Somme, the next Allied offensive was planned for Champagne. This was the decision of General Nivelle, successor to Joffre who had been sacrificed as a result of the Somme failure. Nivelle's tactics were to deliver a quick massive blow, a break-through offensive between the Oise and the Aisne. The opening of the offensive was planned for the first days of April. To give his offensive increased punch Nivelle moved 850,000 men into the area; colossal preparations were made. He left nothing to chance.

Nevertheless he was taken by surprise for the Germans carried out a deep withdrawal, over some ten miles. Nivelle had missed his opportunity. Against the advice of most of his generals, however, he said: 'That makes no difference; we will attack just the same.'

On 6 April Poincaré had to arbitrate between the General Staff and the commander-in-chief. His decision: 'Agreed. Attack.'

War, it is said, is much too serious a thing to be left to military men. In the event the politicians showed up no better; the offensive was a failure. Despite an artillery preparation even more violent than that of the previous year, the Allies achieved no result.

The task of the Foreign Legion Regiment, the RMLE, was to seize the village of Auberive as soon as the artillery lifted. Auberive lies on the edge of a birch wood, the trunks of which were still standing in the midst of the general desolation. The village overlooked a salient known as the 'golf course', an area of flat ground swept by German machine guns. For a week, from 10 to 17 April, the artillery pounded the sector but, as the legionaries immediately noticed, missed the 'golf course'. The order to attack was issued at 4.50 a.m. on 17 April.

It was raining. It was one of those wet windy spring days when the ground becomes slippery as an ice-patch, the mud is holding, clothing weighs a man down and the soil turns into a spongy glutinous substance which clings to army boots. The attack moved off nevertheless.

Having been missed by the barrage, the German machine guns immediately scythed down the first wave of the assault. Lieutenant-Colonel Duriez, commanding the regiment, was one of the first to fall. His place was immediately taken by Major Deville.

The attack failed, however, although the RMLE had succeeded in gaining a foothold on the southern part of the 'golf course' and in an angle of the birch wood. The two sides were firing at each other at point-blank range but without much hope of success. None of the positions changed hands.

The next day the rain stopped but it merely gave way to snow. Yet the Legion had to try again. Captain Maire had been ordered to eliminate the 'golf course' salient. With the stolidity of the old legionary he had arranged his artillery support during the night but he had said to the gunners: 'You are not to open fire except on my order.' He had realised that artillery preparation acted as a signal to the Germans.

At the appointed hour his company moved forward. 'My three leading platoons,' he writes, 'charged in to the assault like a coiled spring. As I had ordered, they fanned out and without difficulty reached the first German line, where the sentries were on the watch. How many sentries were there? All I know is that, without losing a moment, we "skewered" them. Suddenly there came a storm of shell-bursts. It was the barrage – but too late! We were through.'[1]

Thanks to his mistrust of the efficacy of artillery preparation Maire had accomplished a rare feat: at the head of his company he had broken through the German lines and by 5.10 a.m. was out of range of German counter-attacks.

Unfortunately he was alone and that day the Legion paid a heavy price for its loyalty to General Nivelle. One after another the battalion commanders were killed – first Captain de Lannurien, then Captain Germann. By the evening of the 19th the 3rd Battalion's strength was down to 275 men.

Nevertheless the RMLE now had a foothold in the enemy lines. They bombed their way up the main communication trench, the principal enemy artery. During the day they 'mopped up' 7000 yards of it. Finally in the evening the regiment entered Auberive – 'Orders carried out.'

Next day came a remarkable exploit by one of the few Germans who had remained with the Legion despite the war. Sergeant-Major Mader (Max-Emmanuel Mader, No. 6 Company, RMLE) had said: '*Legio patria nostra.* What's good

1. *Souvenirs du Colonel Maire* edited by J.-P. Dorian.

for the others is good enough for me.' So for two years Mader had been making war.

Dawn broke on 21 April, a grey cold dawn revealing an area which had been the scene of severe fighting for three days. Trench 67, along which the 2nd Battalion had attacked on the previous evening, was still piled with corpses, friend and foe intermingled in death. Everywhere was mud, ammunition cases, weapons and torn clothing, formless and colourless. All was grey, sodden and horrible.

In a sap trench the survivors of Nos. 6 and 7 Companies were sleeping, overcome with fatigue. They were sleeping despite the smoke from the fires still smouldering beneath the debris, despite the mud, the wet and the cold. They slept without knowing whether the man against whom they leant was dead or alive, a friend or a prisoner.

Halfway along Trench 67, however, Sergeant-Major Mader was still keeping watch. He had observed that the trench ran northwards and that it overlooked a little valley at the bottom of which, he was sure, something interesting was hidden. His fighting instinct seldom deceived him.

He said to the captain: 'I am sure there are some guns 150 yards further on.'

'Perhaps, but between us and them there is a full company with machine guns. If you take it into your head to go down into that valley, you'll never get out again.'

Further down a legionary was also on watch. 'Sergeant-Major,' he called out softly.

'What do you want, Bangerter?'

'There are some of our chaps advancing down the valley. They haven't seen the machine guns and unless someone warns them, they'll all get killed.'

Bangerter had been a lance-corporal for a month. He was quite young, barely eighteen, and had not yet been hardened by the horrors of war.

Mader smiled: 'That soft head of yours will be the death of you.' He glanced in the direction indicated and growled as if to himself: '"Poilus" of the 168th. Why can't they look where they're going.'

'What'll we do?' Bangerter asked. 'Stir up the B . . . sorry, the enemy?'

Just in time the lance-corporal had remembered his sergeant-major's nationality.

Coldly Mader replied: 'If you stir up the Boches, you idiot, you'll stir up the others too and they'll give you a bullet in the head.'

It was a difficult situation for Bangerter to consider. The French were now precisely in the German line of fire.

But Mader knew his job. In a few seconds he had collected ten legionaries and set off. His decision had been quickly taken: there could be no question of firing on the enemy company but he could bomb his way through. He emerged with his ten volunteers and dashed down into the valley at such speed that they had reached dead ground before the machine gunners had had time to adjust their sights.

The company of the 168th Infantry was still unsuspecting. They had reached the foot of the enemy strong-point which they believed to be unoccupied. Above them the enemy was waiting.

It was at this moment that Mader intervened; he rose like the demon-king behind the petrified Germans and cleared the position with hand grenades. Those who were not left in ribbons by the grenades fled towards the 150 mm battery, of which Mader now had a clear view.

He jumped down into the trench, gave a quick shake of the hand to the captain of 168 Infantry whose life he had just saved and rapped out: 'Not a moment to lose. I'm not having those bastards spoiling my fun. Must catch them up before they reach the guns.'

He ran down the trench, seizing a haversack full of grenades from one of the infantrymen as he passed. He dashed off, only a few yards behind the fugitives and quickly gained on them. Despite everything his ten legionaries kept their heads and followed him, though in some disorder.

Now Mader had reached the last fugitives. They had the unhappy idea of jumping into a dug-out. Mader tossed in some grenades in passing and then erupted alone into the middle of the guns. They were guarded only by a few Saxons, half-awake and waiting for their relief. Still using his grenades, Mader made short work of them.

Other gunners eventually appeared but the timely arrival of the ten legionaries definitely decided the issue as far as the Saxons – and their guns – were concerned.

Result of this action: one French company saved, one enemy company routed, one heavy battery captured . . . and for Mader, the Legion of Honour.[1]

By August 1917 the Foreign Legion Regiment was at Verdun. On 20 August it attacked Cumières Wood, of unhappy memory. The legionaries seemed to have caught the mood of the moment and no longer sang their traditional marching songs, whether serious or sad. They hummed 'La Madelon'.

'Father' Rollet had just arrived to command the regiment. He was still dressed in his khaki linen tunic and according to the 'old hands', who persisted in calling him the 'rope-soled captain', was still shirtless. He had returned to the Legion and for once the regiment was to be in luck.

In the space of two hours Cumières Wood and the Des Forges communication trench, both said to be impregnable – the French had been breaking their teeth on them for a whole year – had been captured. But Rollet did not stop there. 'Since we're having a good day,' he said, 'let's go on.'

1. Sergeant-Major Mader's citation was signed by Marshal Pétain himself. It read: 'An NCO of remarkable courage and energy. An exceptional platoon commander and a genuine leader of men. In the fighting of 17–21 April 1917 he was always at the head of his men and conducted himself admirably. By adroit manoeuvres and accurate fire from his light machine guns he and his platoon achieved the capture of an enemy battery, putting its supporting infantry company to flight.'

Straight away, without drawing breath, he drew up a plan for a second operation. This was nothing less than the capture of an extremely heavily defended strong-point – Point 265 on the 'Côte de l'Oie' [Goose Hill]. Seen from Cumières Wood, the strong-point appeared less formidable.

Having heard of the Legion's initial success, Generals Pétain and Pershing arrived and the 1st Battalion made its assault under their very eyes. In an hour the strong-point was captured. Meanwhile the 2nd Battalion under Major Waddel seized the Côte de l'Oie. During the day the Legion had captured a front of over two thousand yards in a sector where other élite units had been struggling in vain. It had taken 680 prisoners including 20 officers. This exploit earned the 1st Battalion a mention in despatches.

Next day the 3rd Battalion under Major Giudicelli, the 'loud-mouthed' officer, took over defence of the positions. On 4 September the Legion was relieved. Its list of captures in addition to the prisoners included: four 105 mm guns, one 38 mm gun, ten 77 mm guns and thirteen machine guns. Prisoner interrogations showed that it had been faced by four élite German regiments – the 20th, 24th, 221st and 223rd. It could now go back to rest.

The rest was relatively short-lived, however, for the regiment was plunged straight into preparations for the parade at which its colours were to be decorated with the Legion of Honour. The citation was flattering.

> A marvellous regiment inspired with hatred of the enemy and the highest spirit of self-sacrifice.
>
> In Artois on 9 May 1915 under the orders of Lieutenant-Colonel Cot it assaulted the Ouvrages Blancs, breaking straight through all enemy organisations, carrying Hill 104 and pushing forward to Carency-en-Souchez.
>
> In Champagne on 25 September 1915 under the orders of Colonel Lecomte-Denis and then of Major Rozet captured the Wagram strong-point north of Souain:

On 28 September under the orders of Lieutenant-Colonel Cot overpowered strong defences and pushed forward to the trenches and woods of Navarin Farm which it captured.

On the Somme on 4 July 1916 under the orders of Colonel Cot crossed an open space of 800 yards swept by machine guns, took Belloy-en-Santerre with the bayonet and defended it despite intense enemy bombardment and violent and repeated counter-attacks.

In Champagne on 17 April 1917 in front of Moronvilliers Hills under the orders of Lieutenant-Colonel Duriez and then of Major Deville attacked a determined enemy three times its number. In uninterrupted hand-to-hand fighting lasting five days and five nights captured the 'golf-course' trenches and assisted in forcing the enemy to evacuate the village of Auberive by taking him in the rear.

At Verdun on 20 August 1917 under the orders of Lieutenant-Colonel Rollet captured the village of Cumières and its wood, moving with such dash that it overshot the objective assigned to it. The regiment then seized the Côte de l'Oie and Regnéville.

This citation was awarded on 27 September 1917.

On 11 November 1918 came the Armistice. For a few months the Legion formed part of the occupation forces in Germany; then, in small detachments, it returned to Sidi-bel-Abbès.

There it found the German legionaries whom it had left behind in 1914. They were not bitter, nor had they been bored. For four years, like their forbears of 1870, they had 'held on'.

One day in 1917 a party of American journalists travelling in Algeria stopped in astonishment in front of a bar. From within, loud and clear, came the words of a well-known German song, immutable and rhythmic: '*Lieb Vaterland, magst ruhig sein . . .*'

Dumbfounded, one of the journalists asked: 'But isn't that the "Watch on the Rhine", the song of the pan-Germans?'

'Yes,' the Foreign Legion lieutenant replied coolly. 'Why not?'

'Would they be German prisoners?'

'Oh, no; they're legionaries.'

'And with songs like that being sung, you have no fear about them creating trouble in the rear areas?'

It was the officer's turn to be surprised. 'No. Why should we? Here in the Legion the words "loyalty" and "honour" mean something. These Germans are doing their duty as legionaries. They are serving the flag they have chosen. Let them sing what they like.'

It was true. The Germans of Algeria and Morocco had done their duty well. They returned intact to the survivors of the Great War the area entrusted to them.

Messifrê, Syria

16 September 1925

'WHICH OF you can ride?'

In their stiff white uniforms, army hair-cut beneath their *képis*, the young recruits, the 150 men of the training platoon, stayed mute. They had had this game played on them already – 'Who knows how to ride a bicycle?' and the simpletons who stepped forward found themselves on potato-peeling fatigue: 'Who wants a pass?' and the naive ones would be on brothel patrol. Not a soul moved.

The sergeant screwed up his eyes. His voice changed, went down half an octave and took on an air of menace as he asked once more: 'I would like to know if there are any legionaries among you who know how to ride a horse.'

One, then two, then three legionaries stepped forward. They took one pace to the front, stood to attention and saluted. They looked a little too mature for 'young' legionaries. Their faces were tanned, their beards flecked with white. They stood there stiffly, still not sure whether they had not been 'had' once more.

'Legionary Buriachev.'

'Legionary Kussack.'

'Legionary Elisseiev.'

'Where did you learn to ride?'

The same laconic response came from all three: 'Denikin.'

The name Denikin signified the White Russian army which had fought against the Bolsheviks for three years. It had fought practically everywhere – from Kiel to the Ukraine, from Moscow to Vladivostok. Denikin, Wrangel and Kolchak were the names of the vanquished generals whose troops had swept across Europe on horseback and then scattered into exile.

Many of them had foregathered in the Foreign Legion. They had had one final fling in Marseille, sold their medals and their uniforms and had arrived in civilian clothes, nameless.

The story is told that one day, during an inspection at Sidi-bel-Abbès a colonel was struck by the appearance of a 'young' recruit with the moustache of a Tartar and the beard of a prophet. The colonel asked: 'What were you before you came here?'

'I was a general, *mon colonel*.'

The story cannot be confirmed. What is certain is that, disregarding the strict discipline of the Legion and in the secrecy of the barrack room, a certain sergeant would come each evening to kiss the hand of a simple legionary and address him according to the ancient Cossack tradition as: 'Monsieur le *sotnik*' (a Cossack rank equivalent to captain).

The sergeant had previously been his batman.

Following the three Russian legionaries other volunteers stepped forward – other Russians but also Germans, survivors from the old Imperial cavalry – the whole European world of the horse, in fact. In a matter of a few days sufficient horsemen had been found to form the first squadron of what was later to be the 1st Foreign Legion Cavalry Regiment.

Officers arrived from Saumur which had hitherto had nothing to do with the Legion since it had recruited infantrymen only. They brought with them the cavalry spirit and a certain panache. Soon all combined to form a strange mixture partaking of the spirit of Mother Russia, the traditions of Saumur and the discipline of the Legion. Very soon too the 1st Foreign Legion Cavalry gave itself an unofficial title: the *Royal Étranger* ['Foreign Royals']. Quite seriously the regiment maintained that it was the spiritual heir of the old Royal Regiment and could trace its formation back to the year 1635. For good measure it took as its motto that of the

Kings of France: *nec pluribus impar* – in all modesty! The 2nd Foreign Legion Cavalry, formed a few years later, had to fall back on the title *Dauphin Étranger.*

Gradually the cavalry legionaries (they were not referred to as 'mounted' – that term was restricted to the infantry companies which had one mule per two men) adopted a sort of hybrid uniform. The tunic and epaulettes were regulation but the buttons were silver instead of gilt, leggings were worn in place of puttees and the cavalry sabre replaced the bayonet.

Training went ahead at speed. Headquarters had decided to form a horsed regiment, not to look smart on parade but to fight the Moroccan rebels. The Legion required its own squadrons so that it would not be dependent on the spahis.

Two years later the 1st Foreign Legion Cavalry departed for Tunisia. Its task was to pacify the south where certain rebel tribes from Tripolitania had been making trouble.

At Gabès the legionaries quickly became masters of the place, overshadowing even the colonial troops, the tirailleurs or chasseurs. There were inevitable clashes, for Denikin's Russians had the reputation of being ill-humoured and quick-tempered.

There were frequent brawls, principally in the red-light district where brothels were two a penny. The garrison commander was forced to confine the legionaries to barracks except on certain days, but of course the Russians took no notice. Moreover they did not make love like sugar-daddies; they liked to wet their whistles first and then finish up with a good scrap. Many stories of these epic brawls went the rounds of the officers' club. The Military Police never intervened. As if by accident on days when the Russians would be on the loose – particularly pay-days and on their return from operations – the police would be called upon to keep order outside the town.

One day, however, the police put two legionaries inside. This was a real insult since the Legion liked to settle such affairs itself. The captain commanding the squadron to which the

. Legionary at Sidi-bel-Abbès.

10. The 'crazy years' of fighting the Berbers in Morocco. A squadron of 1st Foreign Legion Cavalry in 1925.

. After the battle — a battalion of the Legion comes down from Djebel Sagho, the
.t Berger stronghold in 1933.

12. Scouts of the 3rd Foreign Legion crossing a stream in Cochin China on a tree trunk with the help of a rope — the 'monkey bridge' technique.

13. If there is no bridge the river must be forded.

14. Supply by helicopter on the Hodna during Operation 'Etincelles' (Sparks).

3rd Motorised Sahara Company on parade at Laghouat.

The haul.

16. Crypt of the Legion's museum at Aubagne where colours of the old units are laid up. Captain Danjou's artificial hand in the centre.

two men belonged went to the Deputy Prefect. Their interview was a short one.

'I am leaving on patrol at midday,' the captain explained, 'and I hope to be at full strength. If I am not I shall be compelled to come along with my squadron to refill my ranks.'

The Deputy Prefect got the point. The legionaries were back half an hour later. Their freedom was short-lived, however; they collected 'fifteen over eight' (fifteen days' detention of which eight were in close arrest). Reason: 'Late for morning roll-call.'

In the 'regular army' they would have been run in for 'drunkenness and conduct to the prejudice. . . .' But the Legion had its principles: anything that happened outside duty hours was none of its concern.

With such men the 1st Foreign Legion Cavalry inevitably made an impact. Certain envious souls hoped, though they took care not to say so out loud, that some 'real men' would be found to take the Legion down a peg.

At one moment it was thought that these 'real men' had been found in the shape of the 'Joyeux', the old prison lags who had been salted away in the African Penal Battalions where prison-standard discipline was the rule. With the solidarity customary in the underworld these uniformed hoodlums one day decided that the streets and brothels were their property. They had not the courage to launch a direct attack on the bars frequented by the 'Cossacks' and so they decided on intimidation tactics. Working in teams of six, they would watch for a lone legionary and then beat him up. This happened once or twice.

But the Russians' patience was short and they struck back. Next time came a shout: 'Round me, the Legion!' In a dark street a single legionary was facing six 'Joyeux', belts wrapped round fists with buckles flying. Some of them were carrying broken bottles. White *képis* surged out of every doorway. In a few seconds the street was cleared. Four infantrymen were left lying on the ground; the others departed at full speed.

The next day the same scenario was re-enacted, but with a difference – a 'Joyeux' slipped a live grenade into his opponent's pocket. It went off, tearing the unfortunate man apart. One man dead.

Blind rage gripped the man's comrades. Disregarding their orders – the regiment had immediately been confined to barracks – the legionaries descended on the red-light district. 'Go to it, boys,' one of them said.

The town was in panic. This was a real assault. Two hundred legionaries were on the war-path, but they were unarmed; they carried only pick-helves or iron-shod clubs.

The 'Joyeux' had not taken to their heels. Whether out of fear or as a provocation they had collected their comrades and decided to give battle. They all carried knives.

For two hours the battle raged through the slums, the brothels and the alleyways in a terrifying silence. The legionaries struck without uttering a word. Only the 'Joyeux' could sometimes be heard calling for help.

Everything was in shreds, everything broken. The civilians went to ground in terror and the police, when called upon to help, did not budge. In two hours the Legion had cleared the town.

Casualties among the 'Joyeux' were six dead and twenty wounded. For a long time thereafter Gabès left the legionaries in peace.

The following year they left for the Middle East. The League of Nations had given to France the Mandate over Syria and Lebanon after the departure of the British. This, of course, was not to the liking of the Druses, a fierce mountain people and fanatical Moslems; they looked upon the arrival of the French as a fresh crusader conquest.

A revolt broke out in 1925 starting with the massacre of the

Christian population in south-east Lebanon, and the Legion was ordered to go to the assistance of General Sarrail, Army Commander in the Middle East. The expeditionary force included the 4th Foreign Legion Cavalry and the 5th Battalion of the 4th Foreign Legion Infantry.

Initially the legionaries did no more than patrol suspect villages on which fines had been levied. If the fines were paid, the villagers were held to have 'repented' and to have been 'won over'. If they refused, they were driven out of the village which was then set on fire. In 1925 the theory was that 'the Arabs only respected force' and in fact pacification soon spread like an oil slick.

One day, however, the little post of Es Suweidiya on the flanks of the Jebel ed Druz was surrounded by a large concentration of rebels. No one could get out and any reinforcement was bound to take time. For a whole month the only relief brought to the besieged garrison was by daily air action. The air force bombed the besiegers and, for the first time in military history the parachute was used to deliver ammunition, food and water to the post.

Early in September 1925 the Foreign Legion was ordered to leave its camp at Gazalé and advance towards Es Suweidiya to relieve it. On 25 September the legionaries reached Messifré.

At this point a message reached Major Kratzert commanding the 5th Battalion: 'Five thousand Druses have left Suweidiya to bar your road.' Like the old legionary that he was, Kratzert took immediate action: 'Occupy defensive positions.'

Messifré is a little mountain village grouped around its mosque. There are two little streets with terraced flat-roofed stone houses. Kratzert positioned his cavalry squadron under Captain Landriau in the centre of the village as counter-attack reserve. Outside the village he laid out four strong-points, each to be held by an infantry company.

'Man the walls.' The legionaries had only been there ten minutes but they knew what this order meant; they had to construct a parapet to protect them against imminent attack. From the plateau which sloped gently down to a series of gullies they collected blocks of stone which they transported and assembled. In this way a loose stone wall, some two feet high, was built. By nightfall each company had made for itself an enclosure about a hundred yards square, a sort of fort resembling either a corral on a ranch or an enclosed garden in some Latin country.

The heat was stifling and the legionaries had only very little water. The supply provided slightly less than a gallon per man, of which the kitchen took half for soup- and coffee-making. One of the survivors of Messifré, an American named Doty, writes:[1] 'We lived in piggery, all the more since we wasted half of the water remaining to us by trying to filter it.'

Having established their camp, the legionaries then had to construct an air landing strip. And night was falling. Doty continues:

> At 3 a.m. a sergeant making his round of the outposts thought he heard a stone rattle on the slope before the camp. He put his ear to the ground and heard a rumour of rolling stones and the grunting of camels. '*Aux armes!*' he bellowed. The cry was taken up and passed along; we rose to our feet at our créneaux and were ready in the black and bitterly cold night. Nothing happened for a time.

Suddenly, by the flare of a Very light, the legionaries saw an astonishing spectacle. The entire slope of the hills and the plateau ahead was covered with an avalanche of white burnous. Having been disclosed by the Very light, the Druses – there were about 3000 of them that night – attacked in close order.

Fighting during the night was confused. The infantry company strong-points held out but the Druses had sent suicide volunteers into the village and they massacred the horses and

1. Doty: *The Legion of the Damned.*

their Cossack escort. The survivors withdrew into the houses, the roofs being then immediately occupied by the enemy.

Dawn broke and the attackers withdrew into the shelter of the gullies to draw breath. Then, as soon as the sun was up, they attacked again. '*Aux armes!*' The legionaries had remained at their loopholes, taking advantage of the lull to smoke a cigarette or suck a piece of chocolate, the only sustenance which did not need cooking.

> The sight that now met us [Doty continues] was an extraordinary one and one that might have been terrifying. Down the slope at the bottom of the hill the Druses were pouring down upon us as if it were the very earth itself moving in a landslide – a flood of Druses yelling '*Yah Illah Lah*', 5000 of them on horse and foot. The infantry was ahead; right behind was the cavalry ready to charge through the infantry's opening ranks. They were all in light fighting equipment without their burnous, in their short jackets, their coloured headdresses flapping. They brandished their weapons and howled as they came and at their head were four great black banners flanking the Emir.
>
> For a moment I thought I was dreaming. For riding ahead between their black banners was their Emir on his beautiful blooded stallion and he wore a suit of chain armour and one of those helmets with flaps of chain mail such as you see in museums. Our machine guns were chattering, our rifles were cracking but this mass of men was coming along seemingly unimpeded, coming in fanatical charge yelling '*Yah Illah Lah*'. . . .
>
> In between I had the time to catch a glimpse of a strange and magnificent sight . . . In a rush like that of a wave the Emir topped the mound in the centre of the village with his four great black banners. He dragged his horse upon its haunches, pivoted it till he faced us and simultaneously his guard planted the banners on the mound, sticking them into the soil with great strokes as though they meant them to

stay there for ever. Dismounting calmly under the flapping ensigns, he set himself to directing the fighting from there. Suddenly all the flat roofs of the village were alive with snipers who poured a fire into our camp which we could not answer, busy as we were with the charges down there.

They were still coming, wave upon wave . . . They came down that slope flowing from rock to rock, from wall to wall like a great cataract of water. Then they'd rush the last space and almost reach our wall. Some charged unarmed; we found the body of one with both hands on the wall. He had seemingly charged our machine guns with bare hands.

It was now 9 a.m. The attack had lasted uninterruptedly since 3. The Druses were at the barbed wire, only twenty yards away, within grenade range. Doty goes on: 'Esquizert, a Pole, was responsible for keeping us supplied. He crawled up to me. I stole the whole full haversack, whereupon he said like a schoolboy: "If you don't give me back my grenades, I'll tell the lieutenant".'

At 10 a.m. the Druses fell back. Was it a withdrawal or a manoeuvre? the legionaries wondered. To the right fighting was still going on in the village round the houses where battalion headquarters and the Cossacks were besieged. Doty takes up the story again:

Then someone said, 'Let's get the Emir.' His horse had been led away but he was still standing on the mound, surrounded by his guard, the big black banners flapping above him. I suppose it is not etiquette for an Emir's guard to seek cover. Under well-aimed fire from Fleury, a little man from Montmartre, Constanovich, and several other sharpshooters, they began to drop one after the other. Still they remained, the Emir in his chain armour and casque. Then finally we got him. He was standing one moment, he was down the next and there was nothing on the mound but sprawling bodies and the upright flapping black banners.

On this occasion the relief column arrived in time. Elsewhere the French were not always so lucky; the 'Michaud column', for instance, was cut to pieces at the gates of the Es Suweidiya post; all that was found by the legionaries two months later was a pile of skeletons among the rifled baggage.

Es Suweidiya was evacuated but another celebrated battle took place at Rasheiya. There a column established itself in the old Crusader castle but was surprised to find itself being attacked, not only frontally but through the maze of underground passages, of which it knew nothing. The Druses erupted, apparently out of the ground, and hamstrung the horses of the Foreign Legion cavalry squadron.

These battles gave rise to a song which officers and NCOs of the 1st Foreign Legion Cavalry invariably sing at a guest night standing to attention:

Une colonne de la Légion Étrangère
S'avance dans le bled en Syrie.
La tête de la colonne est formée
Par le Premier Étranger de Cavalerie

Le Druse s'avance à la bataille.
En avant, légionnaires, à l'ennemi
Le plus brave au combat comme toujours
C'est le Premier Étranger de Cavalerie.

A column of the Foreign Legion
Advances through the Syrian desert.
The head of the column consists
Of the First Foreign Legion Cavalry.

The Druses advance to battle.
Forward, legionaries, at the enemy
As always the bravest in battle
Is the First Foreign Legion Cavalry.

By the following year Syria had succumbed and the Legion returned to Morocco. It continued with pacification tasks, opening roads, constructing posts, all with some strategic purpose.

This was the period when the famous Foum el Zabel tunnel was driven between Midelt and Erfoud. In six months a tunnel was driven through a mountain, straight through the rock. The legionaries left a laconic inscription which, although with an element of false modesty, also betrayed their immense pride in their achievement:

> The mountain stood in our way.
> We were ordered to proceed nevertheless.
> The Legion carried out the order.

Pride was visible in the eyes of the legionaries who posed, grouped round their officer when the work was finished. Though apparently relaxed, they were clearly conscious of their feat.

This was the period when the Legion worked with a pick in one hand and a rifle in the other, for operations were still in progress in the Upper Atlas and the whole area was insecure.

In August 1933 the Upper Atlas was the scene of the last of the Legion's major military operations in Morocco before World War II.

Above Ouarzazate rose a jumbled reddish massif where no white man had ever set foot. The Berbers who lived there equally knew nothing of Europeans and sincerely believed that they were ogres who ate children. It only needed an incident for the French to decide to occupy the area.

The incident took place in April 1933. A number of light reconnaissance aircraft flew over the area in order to map it for the Army Geographical Service. They were old Potez fighters, unreliable and vulnerable. One day the Berbers fired on two of them and the pilots, who were not killed when the aircraft came down, were tortured and then murdered. This was taken as a pretext for 'pacification' of the area.

In July four columns concentrated at the foot of the Atlas. From Dades, Todra, Gheris and Assif Melloul, the peaks surrounding Djebel Baddou in the heart of Upper Atlas, they converged on the Berbers who had taken refuge on their mountain. Djebel Baddou stands almost alone, rising proud and forbidding, like a fortress. It is covered with snow in winter and parched in summer. There is no source of water. It was here that, from 5 to 22 August, the last independent Berbers underwent the most terrible siege in the history of pacification.

The warriors who had barricaded themselves on their mountain were no kid-glove fighters; suspecting that certain villages in the valleys had provided guides for the French, their first action was to exterminate all the inhabitants. Between them and the columns concentrated at the foot of the Djebel Baddou there was henceforth nothing but an empty plain from which all life had disappeared.

The attack began on 5 August. The legionaries climbed up rocky faces and crawled along the edge of precipitous cliffs. Once having passed the 6000 ft mark (Djebel Baddou rises to nearly 12,000 feet), the cold became intense even in summer.

The Berbers fought back, taking cover in every tiny cleft in the rocks. Often the legionaries were fighting with rifle in one hand with the other clinging to a rock, and a precipice behind them.

Air support, though available, proved ineffective; the small 100 lb bombs could do nothing against men crouched in caves or sheltering behind slabs of rock. The artillery, whose duty it was to support the infantry from the plain below, achieved no better results. This was a battle of man against man in which every individual had to rely on his own cunning, courage and agility.

There were two to three thousand Berbers on the mountain

– no one could say exactly how many. With them were their wives, children and flocks of sheep which the attackers could sometimes hear bleating from thirst. Sometimes the Berbers would shout offering money for a bottle of water. Some of the soldiers accepted, whether out of pity or avarice. They put down a bottle and moved away. When they returned, the bottle had vanished and in its place was a silver piece.

The siege and the slow advance lasted one week and then a second week. Clinging to the rocks, the French climbed up with the pertinacity of ants. There was never a general engagement, rather a series of skirmishes by a section of infantrymen or even fewer. Firing was continuous but never intense.

Sometimes a certain tribe would agree to surrender. After a short period of negotiation conducted by a French officer of the Native Affairs Bureau, the deal would be concluded. A caravan would then descend the slope to the water points under the watchful eye of the machine gun companies. Headquarters was resolute : by one means or another the besieged would be forced to surrender. Thirst was a weapon, perhaps the most effective.

The Berber chief was called Ouskounti. He was savagely anti-French and determined to die with his men rather than surrender. He must therefore be betrayed – and in fact he was. Secret contact was made between certain members of his tribe and French officers. The Berbers promised to take action. Two days went by. 'Chief Ouskounti will surrender in five days' time,' they promised.

It was a long wait but the French did not release their stranglehold. The Legion pushed its advanced posts to the very summit of the mountain with orders not to fight; they were to rely on surprise.

The five days were over and that night Ouskounti launched his last despairing attack. He poured his warriors through the lines, towards the water points and beyond, towards the plain where they could disperse. But then the legionaries rose. To the sound of the bugle they counter-attacked while the artillery, now on the *qui vive*, blocked the exits.

Resistance ended. The siege had lasted three weeks. During the final struggle the legionaries had worked their way forward in three hours of hand-to-hand fighting, their backs to precipices, sometimes balancing on the edge of ravines 300 feet deep. Using the bayonet, the rifle and the grenade, they had penetrated to the entrances of the caves where the women, the children and the flocks of sheep were cowering.

About midday on 26 August an astounding noise was heard from the mountain. It came from the flocks of sheep leaving the caves to rush down towards the water points, a vast bleating sound, echoing from rock to rock, drowning everything. In a furious cavalcade sheep and goats charged towards the plain, brushing their herdsmen aside, half wild, crazy with thirst. They found the water and plunged in, burying themselves in it, fighting, crushing, trampling. The screen of Legion machine gunners looked on thunderstruck. They would never have believed that animals as docile as sheep could suddenly become so violent. This was stark madness. And a madness which seized the Berbers too. Running behind their flocks, they plunged into the wadi, using feet, sticks and fingernails to claw their way through the animals. They wallowed in the now foul water, some drowning, trampled under foot by the mob of human beings and animals.

The legionaries stood with open mouths, shattered by the scene. They did not dare interfere since they would certainly have been lynched by this thirst-crazy mob. They knew what tortures of thirst must have been suffered by these men who had been deprived of water for three weeks, a record which gives some indication of the hell through which they must have passed and the severity of the discipline to which they had been subjected. They looked more like living skeletons than human beings.

Next day the surrender ceremony took place. In little groups the Berbers came to throw their weapons at the feet of General

Huré, commanding the operation. By midday all had laid down their arms.

The ceremony ended with a march-past of the victors. As the Legion went by the older warriors raised their heads and said, 'It is no dishonour to be beaten by you.'

Part Three

Anne-Marie

1940–19 . . .

Anne-Marie

Song of the 3rd Foreign Legion

Mein Name ist Anne-Marie
Ein jeder kennt mich schon
Ich bin ja die Tochter vom ganzen Bataillon (bis)

I

Mein Regiment, mein Heimatland
Mein Mutter hab' ich nie gekannt
Mein Vater starb schon früh im Feld, ja Feld
Ich bin allein auf dieser Welt (bis)

II

Anne-Marie, das ist mein Nam'
Den ich vom Regiment bekam
Mein ganzes Leben lasse ich, ja ich
Für's Bataillon da sterbe ich (bis)

III

Wenn's Regiment früh ausmarschiert
Der Tambur seine Trommel rührt
Tausch ich mit keiner Fürstin nicht, ja nicht
Sie lebt nicht glücklicher als ich (bis)

IV

Ein Offizier den mag ich nicht
Weil er den Mädchen viel verspricht
Ein Legionär nur soll es sein, ja sein
Ihm schenke ich mein Herz allein (*bis*)

My name is Anne-Marie,
Everyone knows who I am,
For I am the daughter of the whole battalion. (repeat)

I

My regiment is my home,
My mother I have never known,
My father fell early on in the field, yes, in the field,
I am alone in the world. (repeat)

II

Anne-Marie, that is my name,
The name which the regiment gave me.
I devote my whole life to it, I do
For the battalion I would die. (repeat)

III

When the regiment marches out early
With the drummer beating his drum,
I would not change places with any duchess, I wouldn't,
Her life is no happier than mine. (repeat)

IV

I want no officer
Because he promises his girl a lot.
It can only be a legionary
To whom I will give my heart. (repeat)

Narvik

15 May – 7 June 1940

'CONFERENCE OF all officers in the second-class dining-room.'

Everyone was seated except the 'Old Man'. As was his habit, he was walking up and down the central gangway, thumbs in belt. He was small but powerfully built. His lurid language concealed a heart of gold. He was not yet called Monclar but simply Magrin-Verneret. He commanded the 13th Half-Brigade of the Foreign Legion.

'Should we have to abandon ship, there will be no panic. I shall be the last to leave.'

Someone sucked his teeth. The 'Old Man' turned round. 'You don't know how to swim? Neither do I. It's of no importance.' And then to the company in general: 'As a matter of fact I wish to remind you of my orders. When under fire there will be no withdrawal on any pretext. Is that understood? – on no pretext whatsoever. All right? Thank you, gentlemen.'

That was all. It was 30 April 1940 and the officers and NCOs of the 13th had just been told that at last they were going into battle. In the gangway below Loprici, an old Tyrolese with thirteen years' service, was weeping tears of joy – 'It can't be true; we're going to land – it can't be true.'

On the deck of the *Providence*, one of the two vessels carrying the 13th Foreign Legion Half-Brigade, the legionaries were getting bored. To celebrate Camerone anchored in a Scottish port shrouded in fog gave them the willies. Sokewsky, an ex-miner from Lens and a marksman with his machine gun, was singing sarcastically at the top of his voice: 'Under the burning sun of Africa – Cochin China – Madagascar.'

'If you don't shut your trap, I'll chuck you overboard,' came from Brada, a Hungarian known as 'Mr Brada' because of some vague story that he had been in politics.

The WVS arrived on the quay with their little trolleys, offering their wares to the 'Frenchies'; 'Cup of tea?' 'Haven't you got any wine?'

Everyone laughed. The legionaries would laugh at anything now they knew that they were off to fight.

It had all begun five months before – on 1 December 1939 to be exact. On that day came dramatic news: the Soviet Goliath was attacking the Finnish David. The French and British governments decided in concert to despatch an expeditionary force of 100,000 men to go to the assistance of the Finns.

The Foreign Legion was asked for as first priority, of course; everyone knew that it could adapt itself to any circumstances. Finland was up in the north, so skiers would be sent. First in line were the Chasseurs and then came the Legion. So they only understood the desert? Never mind – they would 'muddle through'.

At Sidi-bel-Abbès, Saida and Marrakech the usual procedure was followed: 'Who knows how to ski?' Volunteers flooded in. Everyone was accepted, even those who maintained, quite seriously, that they were roller-skating champions.

Six weeks later the Foreign Legion Mountain Group was formed. It consisted of two full-strength battalions. Its commander was Colonel Magrin-Verneret. It was christened 13th DBLE [13e Demi-Brigade de la Légion Étrangère – 13th Half-Brigade of the Foreign Legion].

The 13th DBLE left from Fez on 1 February 1940 and arrived in Marseille two days later. It moved to Larzac Camp above Nîmes. Although there were no mountains, it was very cold – excellent for acclimatisation. The equipment that the legionaires drew was adequate – climbing boots, skiing trousers, khaki berets and anoraks.

Meanwhile the French and British governments were wondering how the force was to get to Finland; neither Sweden nor Norway would agree to the transit of troops. The problem, however, solved itself; on 14 March Finland, on her knees, accepted an armistice.

Among the volunteers for the 'frozen North' the result was consternation. 'We shall look nice in Sidi-bel-Abbès in our Canadian trappers' uniforms,' the legionaries grumbled.

On the Alsace front, in the Ardennes and at the foot of the Vosges the army was bogged down in the 'phoney war'. The 'sitzkrieg' as opposed to 'blitzkrieg' (Guderian's phrase) had no attractions for the 13th DBLE. It waited. But it went on preparing for battle. Since Finland was now no more, it devoted itself to mountain training. It moved to La Valbonne camp where it learnt climbing, rope-work and various other techniques.

In April 1940 Germany invaded Norway. Still burning with the desire to 'get to it', the legionaries thought: 'Here's a war for us.' Both the central Europeans and the 'red' Spaniards who had enlisted after the republican defeat, dreamed of taking their revenge on Fascists wherever they might be.

The Legion was ordered to Norway. It embarked at Brest in two vessels, the *General Metzinger* and the *Providence*. They stopped at Greenock in the mouth of the Clyde. It proved to be a prolonged stop, for the French boats had not sufficient coal to reach Norway. They refuelled, but even then they could not go for the *Providence* had not enough water to maintain pressure. In some disgust the British provided ships for their guests. The legionaries travelled in luxury like tourists; accommodation in the *Monarch of Bermuda* was of a royal standard.

'Where are we off to?'

'Who cares; the essential is that we're going,' replied Sakowsky as he counted the ammunition allocated to his light machine gun – 1320 rounds. The legionaries each had seventy-five rounds per rifle.

The North Sea was calm in these first days of May and, lying on deck wearing their life-jackets, the legionaries of the 13th DBLE enjoyed this new-style cruise. They were at sea for five days; the majority had never done such a thing. Only one question worried them: 'Where are we going?'

On 4 May they got the answer – Narvik. This did not get them very far, however, since no one knew where it was. Back in France Paul Reynaud sounded his trumpet-call of victory. 'The iron-ore route is cut.'

On the map Narvik is a little black spot clinging to a mountain, sandwiched between sky and sea, a sort of eagle's nest, not quite a port, not quite in the air. The railway runs round the base of the mountain. Below are the port installations.

The town was full of Germans. They had fortified the place, laid mines in the approaches and brought up guns under protection of the railway tunnels. For the moment there could be no question of landing at Narvik. The French Expeditionary Force, legionaries and chasseurs under the orders of General Béthouard, was to land in a neighbouring fjord some twenty miles away.

The convoy anchored on 5 May. 'Prepare to land!' With packs on back and rifles slung the 13th DBLE began leaving the *Monarch of Bermuda* and climbing into the lighters which would put it ashore. Facing it was the little township of Bjervik. The ships were deep in the Skannland fjord.

At this point the German air force made its appearance. All at once the dive-bombers attacked, dropping bomb after bomb. The sea rose in great geysers, the ships pitched and tossed, the lighters rolled. On the decks the legionaries lay flat; for most of them apart from the Spaniards this was the first time that they had been attacked from the air. Those in the lighters, who had to remain standing, their hearts in their mouths, wished they were elsewhere.

The bombing went on for three hours, but, miraculously, none of the ships was hit. Discomfited by the anti-aircraft fire and hampered by the narrowness and depth of the fjord, the Germans had not been able to give of their best. As soon as British fighters appeared, they turned tail.

'An excellent start,' Magrin-Verneret commented imperturbably. He had refused to put on a Mae West and had watched the attack standing on the gangway. This was a splendid way of meeting his enemy, he thought.

Finally the 13th DBLE was landed. That evening it camped in the vicinity, somewhat disappointed that it had not yet seen Narvik. 'And what about our town?': 'Don't worry; you'll see it soon enough.'

'It's odd,' Sakowsky remarked some hours later after having been on patrol in the neighbouring villages; 'the Norwegians pull long faces at us.'

'Why does that surprise you?' Brada asked. 'Did you think they were going to hail you as a liberator? For the moment they only know one thing – ever since we've arrived they get bombs chucked at them.'

In fact, although the Norwegians looked askance at the Germans, they had no love for the Allied soldiers either. When not in their own country soldiers tend to do damage. Their philosophy is a tough one and, sadly, the fact was borne out yet again in Narvik.

Two days went by. Magrin-Verneret and Béthouard had made their reconnaissances and selected their initial objectives.

'Narvik cannot be attacked unless we hold the intervening positions of Bjervik and Meby,' Magrin-Verneret observed.

'Right,' Béthouard replied. 'They are good jumping-off points for the final assault.'

In fact both these places were north of Narvik, on the other side of a peninsula overlooking the port. Attack from the sea would take less time and so minimise the risks inherent in this type of assault. The operation against Bjervik and Meby was fixed for midnight 12 May.

The objective allotted to the Legion was the little town of Bjervik. Like most Norwegian ports, its wooden houses clung to the mountain some way above the sea. Down below German guns were guarding the sea approaches; up above the mountains blocked all exit. The German garrison was trapped, with no alternative but to fight.

As with any landing the Bjervik operation began with an artillery preparation. British cruisers bombarded the village with their heavy guns as the lighters moved inshore. As the first legionaries set foot on shore it was precisely midnight. Above them fires were raging; the Norwegians had been right to be suspicious.

On the narrow strip of beach the men of 13th DBLE realised why they had been trained in mountain climbing. Between them and the village the sea wall rose sheer and on top of it were the Germans, fighting savagely.

It took the Legion two hours to climb that wall and three to capture the village. The enemy had to be 'winkled out', house by house, like extracting a snail from its shell. But by dawn Bjervik had been taken.

The savagery of the fighting on both sides had infuriated the Spaniards. They hunted the Germans out of the village, forcing them to take to the mountains. At 900 feet came the snow-line and at 1800 feet the temperature was −30°C. Many Germans died of cold and fear. There were many falls too, for the slopes were frozen and crevasses numerous. On their return the Spaniards relieved their feelings by plundering the ruins of Bjervik. They were not to be the only looters of the war.

'Collect everyone,' Magrin-Verneret ordered. Then to his Chief of Staff he explained: 'We've got the upper hand. The

Germans don't know where they are. I suggest we pursue.'

Lefort stepped forward. He looked every inch a mountaineer, square-jawed, head high, eyes clear and hard. He was the lieutenant commanding the motor-cyclist platoon. 'Reconnoitre towards the west; I want information on the situation opposite Narvik.'

Lefort departed. Two hours later he returned, happily optimistic. 'We can take Narvik this evening if we want to. The Germans were expecting a frontal attack and our overland move has taken them by surprise.'

Magrin-Verneret passed the news on to General Béthouard, who, however, replied, 'We are under the orders of the British and they think we ought to wait.'

'It's madness. It will be tougher tomorrow.'

The legionaries of 13 DBLE thought so too. They had come to take Narvik and now time was being wasted.

But orders are orders. What neither General Béthouard nor Colonel Magrin-Verneret knew was that, faced with disaster on the French front which now appeared certain, the British had no wish to disperse their resources. In fact they had decided to halt the operation and the expeditionary force was soon to re-embark.

When Admiral Lord Cork, in overall command of the operation, passed these orders on to Béthouard, the latter was taken aback. 'It is a question of the defence of our country, sir,' the admiral explained and he showed the general an official telegram: 'Your forces are to evacuate Norway as quickly as possible. The reason is that troops, ships and equipment are urgently required for the defence of the United Kingdom.' 'Of course,' Béthouard sighed. 'Priority to one's own preoccupations.'

Admiral Cork, however, had been convinced by the French arguments: it was possible to take Narvik before re-embarking. It might perhaps be a flash in the pan but a victory, even a partial, isolated and unimportant one, is still a victory. And in this month of May the Allies were in sore need of one.

Next day Admiral Cork summoned Béthouard. 'Embarkation is planned for the night 1/2 June. It is now 26 May. We have four days to take Narvik. Are you sure that you can do it in the time?'

Béthouard smiled: 'No problem, Admiral.'

'All right. Get on with it.'

Sakowsky was jubilant. He 'knew'. How he had contrived to be one of the first informed he keeps as a jealously guarded secret. The staff had not been able to keep its secret from the Pole, however.

'Where are we going to attack?' Brada asked.

Sakowsky grinned. 'You'll never guess.'

'I'll bet I won't; they must have kept the nastiest bit for us. I know these people on the Staff; they hand us all the dirtiest jobs.'

The Pole then explained that the task of 13 DBLE was to take the moles of Narvik port by assault – under the fire of the German 77 mm guns. 'It's always the same, isn't it?' Brada observed phlegmatically.

The legionaries assembled at the point of the peninsula facing Narvik across the fjord – only a little arm of the sea to be crossed under fire. Then there was dead ground at the foot of the moles. Finally, however, the assault had to be made direct into the mouths of the guns. 'We'll be laughing; that's all I can tell you,' they said.

Meanwhile the brigade had heard that the operation was merely a 'show-piece', since it would be re-embarking immediately afterwards. The legionaries did not care. They would have their battle.

The night was pitch dark when the British artillery batteries opened fire. The German 77s replied immediately from their positions in the railway tunnels above Narvik, aiming at the infantry who were embarking at the end of the peninsula opposite. As at Bjervik the legionaries were ferried in lighters; there were no casualties and the attack began. It was 4 a.m.

'Hell! Aircraft!' In a sky empty of British planes (the

British airfields were fogbound) the Luftwaffe dive-bombers appeared. Down below, at the foot of the moles, the legionaries hastily left their position, now untenable; they moved to one higher up but scarcely less dangerous, facing hidden machine guns.

The terrain was terrible. It was cold and the moles were frozen and glutinous. Rocked by exploding bombs, the legionaries advanced on all fours, stopping to draw breath and then moving on again. 'And then I shall be told about those trumpet charges all over again,' Sakowsky chuckled, dragging his machine gun after him by its bipod.

The Legion now encountered its first real difficulty. It was pinned on the mole, the exits blocked by automatic weapons. It held its ground without retreating but it could not advance either. The Germans grasped the situation and counter-attacked in order to throw the French back into the sea. They had nothing to fear from the naval guns whose shells were bursting too high, up in the mountain.

At this point Magrin-Verneret intervened. Erect and cane in hand he went right up to the front line, knelt beside a light machine gun and calmly indicated a target; then he got up and moved forward. Behind him movement recommenced. Gradually the enemy counter-attack bogged down and then halted altogether. After a quarter of an hour the Germans withdrew. Narvik fell to the French at 5 p.m.

This time the population began to understand. Seeing how the French had thrown the Germans out of the town, the Norwegians thought them capable of anything. They hoped that, with the help of the Legion, Norway could repel the invader. The 13th DBLE was given a hero's welcome. It did not dare say what it knew: 'We are leaving Narvik in two days' time.' It was a secret, pregnant but undoubtedly necessary. General Béthouard, however, had to play the game to the end. He despatched the Chasseurs Alpins in pursuit of the Germans. The Norwegians thought that this was a victorious offensive but it was only a manoeuvre.

The Legion also took part in the bluff. It was ordered to attack south-eastwards. In fact it was only required to force the Germans back far enough to ensure that they had no time to reorganise before the re-embarkation. The legionaries did it with gusto, advancing several miles with whole companies of Germans disintegrating in front of them.

The German covering troops consisted of Poles conscripted into the Wehrmacht and their faith in Nazism was shaky to say the least. In fact, as subsequent information confirmed, many were thinking only of desertion.

Luftwaffe parachutists, however, were at hand and they had been ordered to halt the French advance on the mountain crests. Foreign Legion versus German parachute troops: their meeting did not take place – not yet – but it was destined to do so four years later at Cassino.

'What are you up to?'

'Orders. We've been told to make dummies.' Brada shrugged his shoulders. This was no work for a legionary.

To deceive the enemy and make him think that the legionaries were still in position, the Staff had had the idea of manning fake positions with dummies stuffed with rags. It was a trick which would work at least until daylight. So, when night fell, the legionaries raided the clothing stocks found in Narvik and started playing dolls.

'Look at my *feldwebel* [German sergeant-major]; isn't he lovely?'

'Not as lovely as my *fähnrich* [German 2nd-Lieutenant], eh, you idiot.' The clothing concerned had been found in the depot of the provident German supply service.

On 6 June, under the eyes of the horrified Norwegians, the 13th DBLE re-embarked at Narvik. To the very end the legionaries kept the secret, saying, 'We're not leaving the country. We're just moving to another position.'

The Norwegians only realised the truth when engineers of

the Royal Navy began blowing up the port installations and distributing food from the captured German depots. For them hope was now dead. Night fell on Norway, a night lasting four years.

In the holds of the cargo ships taking them back to Brest, which they had left only six weeks before, the legionaries were saying nothing. They were obeying orders; that was all.

None of them suspected that they were very soon to be confronted with a choice, a choice between the war and the armistice. Half chose the war and the other half the armistice. They said farewell to each other in London, the first half going to Egypt and the second to Morocco.

The first half advanced from east to west; the second half made the journey in the opposite direction. They came together again three years later, not on opposite sides but shoulder to shoulder – in Tunisia.

Bir Hakeim

24 May – 11 June 1942

'THIS TIME we'll go all the way to Alexandria.'

General Rommel had made up his mind. He was going to push his tanks right forward – as far as the Nile. He was not the only optimist. In Rome the Duce was getting ready the white horse on which he proposed to make his ceremonial entry into Cairo.

When Rommel had been forced to evacuate Benghazi on the west coast of Cyrenaica in the previous December, his men had written on the walls of the town: 'Back soon – in three months.' They had in fact come back only thirty-five days later. From the beginning of 1942 Rommel had got his second wind. He had now been in Africa a year and the complex hesitation waltz of offensives and withdrawals which the British on one side and the Italians and Germans on the other had inflicted upon themselves, was at an end. That had been his resolve.

Already he could see himself on the Nile, on the Suez Canal and even further, pushing his tanks across Palestine and Iraq to join hands with the troops attacking the Russians on the frontiers of the Caucasus.

For the British the stake was vital. Rommel could not be allowed to enter Egypt. That would be the end of their lines of communication, of their supremacy in the Mediterranean, of the war. There was only one essential – to stop Rommel.

For compelling reasons of supply the German offensive, like the British counter-attacks, was tied to the coast. Along it ran a road on which travelled long columns of lorries carrying petrol, food, ammunition and spare parts. The tanks, on the other hand, had the whole desert to manoeuvre in, subject only to the limitation of the 'point of no-return'. They always had to keep sufficient fuel to be able to return – or pursue.

This was what Rommel was thinking after 24 January, the day on which he retook Benghazi. He wanted to advance to Alexandria but, once again, he had to have enough petrol. Malta, the British island in the heart of the Mediterranean, was a thorn in the flesh of German shipping. Rommel was forced to wait three months before he was ready. Now, on 20 May 1942 he was prepared.

'Forward! All the way to Alexandria this time.'

That same day the 1st Free French Brigade in position in the strong-point of Bir Hakeim received a message from British headquarters: 'German offensive imminent. Increase patrolling and lay minefields.'

In his command post General Koenig smiled. 'That's it then. The battle is on.'

The 1st Free French Brigade had been at Bir Hakeim for more than three months, since 14 January to be exact. It had taken over from the New Zealanders who had withdrawn to a similar locality further east, Bir el Gobi. Its task was to cover the southern extremity of the stop-line established by the British to bar the way to Rommel. It was not in fact a line but a series of fortified and entrenched strong-points linked by minefields.

Since Rommel's first objective was Tobruk, the British line ran slightly forward of the town, starting from Gazala on the coast and running south to Bir Hakeim, just under twenty miles away. Bir Hakeim means 'Old Man's Wells'; the Brigade found no old man there nor did it find any wells which must have been buried in the sand centuries before.[1]

1. Some historians spell the name 'Bir Hacheim' which means 'Dog's Wells'; there was no dog there either except one belonging to the Marines. The precise name is unimportant.

This 1st Free French Brigade was a peculiar formation comprising diverse units which had little in common except a desire to fight the Axis. It consisted of three 'colonial' regiments: the 1st Battalion from Ubangui, the 2nd Battalion from the Pacific and the 1st Battalion of Marines with further marines serving the anti-aircraft guns. It also included the 13th DBLE commanded by a non-French officer, Lieutenant-Colonel Amilakvari.

From Narvik to Bir Hakeim the 13th DBLE had come a long way in a year. As a reminder of its northern campaign it had kept only its 'alpine' khaki cloth beret; otherwise it had 'gone British' – shorts and open-neck shirt, puttees and ammunition boots. It had of course kept the blue belt and white *képi* which, contrary to all regulations prescribing a steel helmet, it wore even in battle. In any case the 13th DBLE was at Bir Hakeim and, whether with or without tin hats, it would fight.

The year spent between Narvik and Bir Hakeim had not been wasted. In June 1940 the 13th DBLE had been in Brest hoping to take part in the final battles, but the armistice had been signed and the Brigade re-embarked. By 20 June it was in London.

There a certain General de Gaulle offered the legionaries a choice: 'Either you want to go on fighting and I accept you with pleasure or you accept the armistice and we will send you back to Morocco.' Half of the legionaries chose war; they included the 'red' Spaniards who still had certain accounts to square.

In March 1941 the 14th DBLE (it had temporarily changed its number) arrived in Eritrea. The British raised their eyebrows: 'Who are these people?' Magrin-Verneret, now called Monclar, still as calm and laconic as ever, merely replied: 'You'll see the colour of the money.'

They soon saw. On 28 March the Legion bundled the Italians out of Keren. The British comment was: 'You swept them away like a broom sweeping up dust.'

On 7 April in front of Massowah the Legion was ordered to capture the Italian fort commanding the position. The assault was made at dawn and by midday the fort had been taken with 10,000 prisoners. 'Not too tough?' the British asked. 'The thirst was the worst thing,' Monclar replied.

Thirst was an old crony of the Legion's, however; it was to be present at Bir Hakeim.

Ever since 20 May Rommel had been driving eastwards. His first bound was intended to take him to Tobruk. He decided to attack the British defences frontally at Gazala and make a diversionary attack southwards against the strong-point held by the French. He had a low opinion of them; in 1940 he had been commanding one of Guderian's divisions and had seen French soldiers surrendering. They would do the same now, he thought.

To deal with Bir Hakeim he despatched the Italian Ariete Division. 'That will keep them busy,' he said. 'Meanwhile I will deal with more serious matters.'

At Bir Hakeim the attack was expected on 22 May. By the 23rd nothing had happened; the Italians were stuck in the sand. On the 24th Bir Hakeim became uneasy – was it going to miss the battle? On the 25th Koenig sent out patrols all round. Still nothing. Next day there was still the same vacuum. Bir Hakeim was deserted – 'a desert in the desert' people said. It was true. No one could find the enemy at all.

On the 26th, however, Rommel took matters in hand and reorganised the force destined to attack Bir Hakeim – 'Clean that lot up for me tomorrow.' The same day a French patrol clashed with the enemy advanced guard which was superior both in numbers and equipment. De Lamaze, commanding the reconnaissance, was accompanied by gun-fire all the way back to Bir Hakeim.

Night fell. 'Things are on the move,' Legionary Dammane said to himself; he was on duty at a gun on the east face of the camp and all round him in the desert he could hear the roar of engines and the grinding of tank tracks. It was impressive.

'I think it's for us this time,' he said to the man next him, Corporal Ilia, a Spaniard with the complexion of a ripe olive.

'At last. We'll be able to send them sky-high.'

Morale was clearly good.

When morning came the legionaries could see masses of tanks manoeuvring on the horizon. 'Alert!' but headquarters replied: 'They are British tanks. We have been told that a counter-attack is going in on Gazala which is threatened from the south.'

No one opened fire but headquarters were wrong; it was the Italians. They came on calmly and opened fire at less than six thousand yards. 'Aim! Fire!' The Legion's 75 mm anti-tank guns spoke. Legionary Dammane was in high spirits; he took meticulous aim. 'Fire!' At the end of the trajectory a tank exploded. Dammane had not even realised that he had been wounded. He was to be wounded three times during the siege.

The anti-tank fire was effective and thirty-five tanks were destroyed outside the perimeter. The remaining five which got through were at once attacked by the Spaniards with their war-cry: 'As at Madrid.' They threw bottles of petrol on the engine cowlings and fired bursts through the aiming slits. Seventy-six Italians, including one colonel, surrendered.

Further north a strong-point held by an Indian brigade had fallen during the night without resistance. The Italians handed their prisoners over to General Koenig, saying: 'We have nothing for them either to eat or drink.' Koenig took them over. Everyone's water ration was cut.

Even before the battle, thirst had been the worst enemy at Bir Hakeim. All survivors agree that the absence of water was worse than the savagery of the fighting. The legionaries sucked pebbles and, following the example of the besieged at El Mungar, drank their own urine. But they held on.

On 28 and 29 May the enemy kept up his pressure. His tanks did not advance but they blocked every exit and prevented all supply. Then on the 30th, to everyone's amazement, there was nothing there. The enemy had vanished. 'Where have they gone?' asked the astonished legionaries.

They had left on Rommel's orders. His breakthrough had not gone according to plan; the Italians should have burst open the Bir Hakeim barrier and they were still a long way from doing so. Rommel decided to break off the operation against Gazala to go and deal with the French himself.

On the Allied side it had been realised that the Germans were in difficulties. Their withdrawal from Bir Hakeim should have led the British commanders to begin the pursuit and, if necessary, counter-attack. That, at least, is what Koenig thought. Next day he despatched a battalion to occupy an important cross-tracks near by. The task was allotted to the Battalion from the Pacific.

On 2 June the battalion was attacked by dive-bombers. 'We are losing men,' it reported over the radio, 'and cannot protect ourselves effectively.' 'OK, come back,' was the reply.

That evening Italian officers appeared before one of the obstacles outside the minefield. 'Come on in,' the legionaries shouted. The Italians advanced. Midway through their eyes were blindfolded and they were taken to Koenig's headquarters. They had come to offer honourable surrender to the garrison.

'Otherwise,' they said, 'it will cost you only unnecessary casualties and suffering.'

Koenig was polite but firm; he replied that naturally this was out of the question. '*Grandi soldati*' was the emphatic Italian reaction.

The move clearly heralded an attack. 'Something tells me,' said Sergeant Echstein, 'that what we have had served up to us so far is only the *hors-d'oeuvre*. Now we're going to have fun.' Echstein was a real old soldier of twenty-four years' service; he had been everywhere – Morocco, Tonkin, Madagascar and Narvik. Now he was at Bir Hakeim. On the 27th he had himself destroyed five Italian tanks with his anti-tank gun.

Echstein was not mistaken. One hour after the departure of the Italian envoys 105 mm shells rained down on Bir Hakeim. They went on doing so. For good measure the camp was bombed by some thirty aircraft. Jets of yellowish dust rose in the air, accompanied sometimes by an explosion as a vehicle was hit or an ammunition dump destroyed. The aircraft circled round until evening; the 105s were still firing. One of the first to be wounded was Sergeant Echstein. 'It's nothing,' he said. 'I can still carry on.'

At dawn next day came a lull. A British prisoner captured during a patrol two days earlier arrived in the camp, carrying a call to surrender signed by Rommel in person. Koenig had it translated.

> To the troops of Bir Hakeim.
>
> All further resistance would lead only to useless bloodshed. You would suffer the same fate as the two British brigades located at Got Ualeb which were annihilated the day before yesterday.
>
> We will stop fighting as soon as you hoist the white flag and come over to us unarmed.
>
> Rommel, Colonel-General

Koenig's reply was immediate. He opened fire on all German vehicles within range. Then, without pausing, he dictated an order of the day.

> General order:
>
> 1. We must now expect serious attack combining all resources (aircraft, tanks, artillery, infantry). It will be powerful.
>
> 2. I repeat my orders and my certainty that each man

will do his duty without flinching in the place allotted to him whether cut off from others or not.

3. Our task is to hold at all costs until our victory is assured.

4. Explain this clearly to all, both NCOs and men.

5. Good luck to you all.

Headquarters, 3 June, 9.30 a.m.

Koenig

General Koenig's language was the sort of thing that the legionaries understood and appreciated, all the more so in that from London the BBC was giving a sensational description telling of Bir Hakeim's heroic resistance, characterising it as an epic and putting into General Koenig's mouth the type of statements normally attributed to General Cambronne.

Koenig was infuriated and sent the following message to British headquarters:

> French-language broadcasts from London have effrontery to attribute to me ludicrous replies to the enemy's demands for surrender.
>
> I replied to the first demand firmly but courteously. Others have not been answered. I am a soldier not a comedian. I request that the defence of Bir Hakeim be not sentimentalised.

Rommel's plan required that, under all circumstances, Bir Hakeim should fall not later than 27 May. It was still fighting on 10 June. Even worse from the German point of view was the fact that, apart from certain minor encroachments, the defence was still basically intact. 'What the tanks cannot achieve,' Rommel thought, 'will be done by thirst.' Koenig thought the same way. He took precautions against any weakening and on 9 June sent out an Order of the Day to the garrison:

We have been doing our duty for fourteen days and fourteen nights. I ask that neither officers, NCOs nor men give way to fatigue. With every day that goes by it will become tougher; that will not make the 1st Free French Brigade afraid.

Koenig was a soldier and once more his language went home to the legionaries. They braced themselves. They were ready. The only man not to hear was Sergeant Echstein. He had been wounded yet again that morning and this time it was more serious; a shell had taken off his right hand. (Despite his wound Echstein made the break-out with his comrades. Exhausted but refusing to be helped, he fell behind and was eventually posted missing.)

The British, however, now concluded that the defence of Bir Hakeim was of no further value for the battle which had been lost elsewhere. To prolong the French sacrifice served no good purpose. 'Moreover,' Auchinleck's staff observed, 'these Frenchmen will be much more useful to us for counter-attack.' It was accordingly decided to evacuate.

'All very well,' said Koenig's officers. 'It's fine to evacuate but we have first to break through the encirclement of the position and then fight a delaying action to avoid being surprised in full retreat.'

Calmly Koenig drafted his orders: they were clear; nothing was left to chance. In the evening a detachment of sappers would lift the mines over a 200-yard wide corridor so that units could move out rapidly south-eastwards. The infantry were to lead, make the breakthrough and push the enemy back to give the vehicles time to emerge and get away. 'D-Day' was fixed for 11 June at midnight.

But between orders and their execution a great gulf is always fixed. Everything started badly; the engineer platoon responsible for lifting the mines could only open a narrow passage barely twenty yards wide. There was indescribable chaos, units piling up like a crowd at the door of a popular cinema. The crowd swelled, overflowed, trampled on each other, shouted. Yet strict orders for silence had been issued: the Germans must not suspect anything; any noise could lead to hundreds of casualties. For the same reason vehicles had been ordered not to start their engines until the last moment – 2 a.m. It was hoped that by then the infantry would be through.

It did not work out. At 2 a.m. the vehicles moved off but they merely added to the confusion.

Lieutenant Dewey was an old officer of the Legion. Being too young to enlist in 1918, he had only one thought in his mind – to become a legionary. He commanded a group of five men, the crew of a tracked jeep, the British-made Bren carrier. It was by no means a tank; it was simply a motorised steel box on tracks, neither comfortable nor pretty.

Dewey arrived at the outer barrier. He realised that, unless someone did something, it was all up; it would be daylight in less than two hours and then the entire garrison would be discovered in the open facing the German tanks.

'We must get through.' Dewey cleared a way for his vehicle and then said to his driver: 'Drive quite straight. Understand?' – 'Yes, sir.'

When a legionary drives straight on, it's straight on. The driver stamped on the accelerator and moved off at full speed – 20 m.p.h. He opened the track through the minefield and behind him the infantry drove forward at the double.

Suddenly a German machine gun appeared in front of the Bren carrier. 'Drive on! Straight!' The vehicle drove on, crushing the machine gun and its crew. It was through. It continued the mad chase and emerged between two lines of sandbags to find itself facing a German 77 mm gun. This was too big a mouthful, but never mind – 'Drive on!' To lend force to his shout Dewey had stood up in his seat. Gripping the steel flaps he looked like a general of the old Empire sounding the charge for a formation of lancers. 'Drive on!'

The 77 fired – at point-blank range. The Bren carrier

shuddered and stopped; the shell had gone straight through it. The driver was dead and so was Dewey. But the trail had been blazed. Behind the carrier more legionaries arrived on foot; they overturned the gun and swept its crew away.

The job had been done; the Bir Hakeim garrison was through. It had lost a quarter of its 4000 men, including 500 prisoners taken on the spot or rounded up in the desert.

From the official point of view Bir Hakeim was not a victory. Sieges which end in withdrawal are generally considered lost. Nevertheless, taking an overall view of the war in Africa and the war as a whole, Bir Hakeim undoubtedly carried the weight of a victory.

Rommel had counted upon the position falling on 27 May; it resisted until 11 June. This delay of fifteen days was fatal to the German offensive. The British army was now regrouped on its second line of defence at El Alamein and could no longer be destroyed. Admittedly Rommel was able to take Tobruk but its capture meant nothing. The British had abandoned it.

Rommel continued to advance but when he arrived in front of Alexandria he could go no further. Here was the justification for the fifteen days of Bir Hakeim.

Later Colonel Amilakvari, the officer commanding the Legion at Bir Hakeim, was killed at El Alamein. This was the last concession which the Legion was to make to its enemy. It took part in the pursuit, chasing Rommel back to his starting point – Tunisia.

There, in Tunisia, while Axis soldiers in their thousands were ending their campaigning days in prisoner-of-war camps, the 13th DBLE was at last reunited with its old comrades from Morocco, now forming a Foreign Legion Mobile Regiment. Henceforth they were to fight together, from Italy right through to the Danube.

Phu-Tong-Hoa

25 July 1948

In 1947 the Legion was in Indo-China. To be accurate, it had never left; ever since 8 November 1883 when the first Foreign Legion battalion had landed at Haiphong to serve under Admiral Courbet, there had always been units of the Legion in Tonkin.

There were four independent battalions grouped along the frontier with China from Mong Cay to Lao Cay. A single road connected them – Route Coloniale No. 4. For sixty years its history was that of any ordinary colonial road serving purposes both strategic, economic and touristic. Later its name was to be written in blood in the history of the Legion.

French Indo-China had been a troubled country ever since the end of World War II and the Japanese surrender. Initially it had been handed over to two of the great victor powers, China and Great Britain, one holding the North and the other pacifying the South; gradually, however, the ex-colony was returned to French sovereignty. In the North Thierry d'Argenlieu's soldiers found a Vietminh army serving a pro-Communist government; in the South Leclerc's men clashed with guerrillas who, though apparently disorganised, unarmed and leaderless, were nevertheless ubiquitous.

In the North the Vietminh was preparing for war but had no intention of waging it until it could put full-scale divisions into the field. In the South, on the other hand, the war took the form of latent but permanent terrorism, of constant insecurity; it was a continuing process, spreading gradually from Saigon to the mouth of the delta. Headquarters allotted priority to counter-guerrilla action in Cochin China.

By 1948 the Legion had sent two of its best units to Indo-China: the 13th DBLE, the men of Narvik and Bir Hakeim, still wearing the legendary khaki beret from the Narvik days, and another formation comprising the 3rd Foreign Legion, the 2nd Foreign Legion and the 1st Foreign Legion Cavalry. Over a period of six years the Legion despatched a series of further units including the 5th Foreign Legion, re-formed after withstanding the Japanese attack on the Tonkin garrisons in March 1945, and the remainder of 1st Foreign Legion Cavalry. In 1948 and 1949 came two Foreign Legion Parachute Battalions. In addition there were a number of independent units – engineers who were distributed by companies, transport, supply and welfare units, heavy mortar companies and so forth.

Whenever French headquarters had a special problem to solve or unit to form, it had only to apply to the Legion. The Legion could always produce the necessary 'specialists'. On arrival in Tonkin, for instance, 1st Foreign Legion Cavalry, which was normally equipped with light reconnaissance vehicles, armoured cars and half-tracks, was issued with light amphibious vehicles known as 'Crabs' or 'Alligators'. For 'Operation Castor' (the capture of Dien Bien Phu) General Gilles required a mortar company consisting of 120 parachutists. It was formed within the month.

In Indo-China, as elsewhere, the Legion knew how to 'make out'; very quickly it adapted itself to this new form of war, to the life it necessitated and the hardships it imposed.

The basic organisation in the sectors into which Indo-China was divided, and the essential factor in pacification, was the military post established in the heart of a 'delicate' zone or near a 'suspect' village. It had three purposes: to establish the French presence; to discourage the enemy by frequent patrolling, thus making the country as insecure for him as it had been for the French; finally to reassure the population and so cause them to swing over to the side of France.

A map of the posts in Indo-China would show as many little flags as there are names on a 1 : 1,000,000 map. How many were there of these little islands lost in the bush, in the high mountain valleys or at cross-dikes in the delta? Perhaps five thousand, perhaps ten.

Initially a post was a rectangular enclosure measuring about thirty to forty yards square. Around it would be a rough wall of stones, breeze-blocks or merely stakes driven into the ground. In the centre of the little courtyard was invariably the French flag. This was the geographical centre, the assembly point for all social life; everything went on there – the despatch of patrols, the hand-out of soup, parades, Sunday mass. Beside the symbolic flag pole was the mortar pit. Around the courtyard and constructed against the wall were the buildings, the commander's room, the barrack room, the food store, the ammunition magazine and the armoury. Sometimes there was a dining hall, a sick bay and a 'clink'. That was all.

At each corner stood a blockhouse. Depending on the resources provided by the engineers, it might or might not be roofed over; it was protected by sandbags, corrugated iron, stakes or anything else which the initiative of the men or the commander had been able to find in the area. Everyone's main purpose was to improve this place where they frequently spent two years, their whole tour of duty.

Sometimes the post was overlooked by a tower. Its obvious purpose was to extend the view beyond the immediate edge of the forest, to spot guerrillas moving through the rice-fields or to keep an eye on the neighbouring village.

Initially merely a look-out post, the tower gradually took on the function of the keep in a medieval castle, the ultimate point of resistance. In Cochin China as in Tonkin many a tower in a besieged post held out to the end.

The Legion established its first posts in Cochin China early in 1948. Its sectors were located in the delta, from the Joncs plain in the extreme south of Indo-China to Camau. It was also responsible for the area extending from the gates of Saigon to the Cambodian frontier, along Route Coloniale No. 2 north of the Joncs plain. The 3rd Foreign Legion was in the south and 13th DBLE in the north.

The latter set up its headquarters at Hoc Mon. In a few months it constructed the most remarkable fortress in all Indo-China, a superb complex of red brick providing everything. The legionaries had built the entire headquarters with their own hands; they had drawn up the plans, installed the plumbing, brought in the electricity. The place wanted for nothing. From the outset the Legion did the best that it could for itself. Its numbers had never been higher nor the pay so good ever since its formation. The Legion was rich and was determined to live comfortably without becoming dependent on the Red Cross or other welfare services.

But comfort did not mean that it was not at war. The war was there, all around, and sometimes its waves battered against the little walls of the posts.

Phu-Tong-Hoa is a small village eleven miles from Bac Kan in northern Tonkin, near the famous Route Coloniale No. 4. It is merely a collection of Chinese houses at a cross-roads, a few yards from which, on a little hillock, a post had been established. It was built, like most others in the area, with a perimeter fence of wooden piles, embellished with four concrete towers, one at each corner. It overlooked an area of uncultivated rice-fields encircled by mountains rising gently towards Deo-Giang in the north which reached a height of 3500 feet. The garrison consisted of No. 2 Company of the 3rd Foreign Legion.

Incessant patrolling had been carried out ever since 20 July 1948 and large enemy concentrations had been reported. The post was an obstacle to the movement of Vietminh troops. It had to disappear.

On Sunday 25 July a fine rain was falling, sweeping across the valley and reducing visibility to a few hundred yards. The surrounding hills were still enveloped in the morning mist. Alert orders had been issued on the previous day and the 104 legionaries under Captain Cardinal lived in battle order with their weapons beside them. The day passed without anything happening.

At 7.30 p.m., as it was getting dark, an explosion shook the post. Down in the plain the Vietminh had brought a mortar battery into position unobserved and it was now pounding the post; almost at once it was joined by 75 mm and 37 mm guns. No. 316 Vietminh Division was in action for the first time.

'Take post!' The legionaries jumped to it, every man to his place. At 8 p.m. a well-aimed shot breached the western face of the perimeter. Five minutes later Captain Cardinal, the company commander, was killed. Lieutenant Charlotton, who took his place, was killed in his turn a quarter of an hour later.

'Lieutenant, you are now commanding the company.' At his post in the courtyard 2nd-Lieutenant Bevalot nodded but did not reply. He had only landed in Indo-China a few days before.

He left his post and enquired for news from the radio operator: 'I've alerted Bac Kan. They are going to do all they can.'

'Try again.'

'Blast! These bastards have smashed my aerial.'

Phu-Tong-Hoa was dumb. It was deaf too, which was more serious. Bevalot went round the position encouraging the wounded and then resumed his place in the south-east blockhouse.

'Can you see anything?'

'*Nichts*,' was Lance-Sergeant Huegen's reply.

The sky was pitch-black and there was no moon. Nothing was visible more than twenty yards away. After an hour and a half's artillery preparation the firing stopped at 9 p.m. Instead silence reigned, but the legionaries knew that this was not the end. On the contrary the worst was to come.

To the sound of bugle calls waves of Vietminh advanced. The assault came from three directions aiming for the south-east blockhouse where Bevalot was, and for the two breaches in the western perimeter face. Hand-to-hand fighting went on for half an hour. Enemy corpses piled up but the pressure did not slacken. The tidal wave came on unceasingly.

'Lieutenant, they're inside the post!' Lance-Sergeant Huegen took a light machine gun and went to reinforce the breach, the defenders of which had been killed. For a quarter of an hour he fought alone. Paulin, who went to help him, was killed with a bayonet thrust. At 10 p.m. Huegen was killed in his turn.

Bevalot took stock of the position. The post was swarming with Vietminh. Resistance continued only at two points: the northern blockhouse where Sergeant Guillemaud was besieged but was defending himself with incendiary grenades, and the south-eastern tower, where the survivors were still holding out round their lieutenant.

As soon as the alert had been given a relief column had set out from Bac Kan. But the road was bad and the Vietminh did their best to delay the force by obstacles, road-blocks and ambushes. By 11 p.m. it had covered only four miles. Accordingly in Cao Bang Colonel Simon commanding the 3rd Foreign Legion decided to take action. He had fifty miles to cover. As he left he could not hope to do more than perhaps recover some of the wounded, provided the Vietminh had not finished them off.

The survivors of No. 2 Company fought on. All they could hope to do was to hold out as long as possible. 'You're in luck, Lieutenant. You're beginning your career in the Legion with a Camerone.'

In fact it was a repetition of Camerone. Bevalot knew the story. He was rewriting it. 'Fix bayonets!' He attempted a sortie, drove the Vietminh back across the courtyard and gained contact with the western blockhouse.

As it so happened, at this precise moment the clouds rolled away and the full moon shone out, lighting up the savage struggle. 'There's a bugle!' It was the Vietminh. They fight only in darkness and the moon upset them. They were sounding the retreat. By 1 a.m. the legionaries had cleared up the last enemy-held points.

Phu-Tong-Hoa was three-quarters destroyed but it had held out. At dawn Bevalot ordered the burial of the dead, the captain and twenty-one NCOs and men. He collected the wounded into the dining hall; there were thirty-three of them.

When Colonel Simon arrived forty-eight hours later with his relief column, his first surprise was to see the tricolour flag still floating at the top of its pole. The relief column deployed before entering the post. The colonel got out of his jeep and moved forward.

From the post the salute sounded. Bevalot received his colonel in regulation fashion – the garrison of the police post paraded in uniform with belts, epaulettes and *képis*. The courtyard of the post had been cleaned up. It was as if nothing had happened at Phu-Tong-Hoa.

Meanwhile in Cochin China the guerrilla phase of the war ran its course but the rebels were showing signs of exhaustion. Pursuing its pacification policy, the French army gradually won over the stake for which the Vietminh were fighting – the population.

With a psychological insight of which they might have been thought incapable, the legionaries inspired the villagers with new confidence, rousing their sympathy and gaining their support. They threw themselves into the life of the country, re-established the prestige of the administration, opened schools, roads and markets. They built hospitals and dispensaries. They became amateur administrators, judges, schoolteachers and doctors.

Tuyen Quang

1885–1954

The 'Old Man' was in a bad temper. The legionaries of No. 3 Battalion of the 3rd Foreign Legion kept their heads down; when the 'Old Man' was in a bad mood it was best for everyone to sing small, all the more so in that on this occasion Colonel Charton had every reason to be cross. He was holding the 'hedgehog' of Cao Bang right at the end of Route Coloniale No. 4 (RC4).

Until recent months the garrison, two regiments and three thousand civilians, had been supplied via this road, military convoys starting from Langson and passing through Dong Khe and That Khe. Soon, however, the Chinese-trained Vietminh (the Chinese frontier was only a few miles to the north) had turned the road into 'hell's passage'. The French found the solution; they turned Cao Bang, isolated at the end of its road, into a frontier desert island, a 'hedgehog' protected all round.

The fact that the place was besieged did not affect morale at Cao Bang. Charton was quite satisfied; his strong-point had been constructed entirely by the Legion and was therefore solidly built; he estimated that, to attack it, Vo N'guyen Giap, the Vietminh general, would have to sacrifice twenty thousand men, something he could not afford. With air supply Cao Bang, only half an hour from Hanoi, was an adequate 'hedgehog'.

No one either in Saigon or Hanoi, however, shared Colonel Charton's optimism. By this month of September 1950 China had passed under the control of the Reds and they had stepped up the training of the Vietminh divisions to the maximum.

An offensive was imminent. It would be tough and woe to any garrison, whatever its strength, which might stand in its path.

General Alessandri had commanded the French column which escaped from the Japanese in March 1945 and fought its way for five hundred miles from Tong on the delta into China passing via Dien Bien Phu and Lai Chau in the mountains. Now he wished to stand and fight the Vietminh. 'Let us take advantage of the weakness of their divisions to attack them,' he said. 'Wherever we can strike, we should do so.'

General Carpentier, his superior, took a different view, however. 'We shall never be able to hold out. We must evacuate the line of garrisons along the frontier, Dong Khe, That Khe and Cao Bang.'

Alessandri fought for his concept, but Carpentier would not change his mind. Finally seniority carried the day and on 15 September the order reached Charton: 'Evacuate Cao Bang.'

This was why, on this 20 September 1950, Colonel Charton, the 'God' of his men, the born fighter who had never given way to anyone, was in a bad temper. He had been ordered to pack up in face of the Vietminh without even being allowed time to fight them.

What Charton did not know, at least not yet, was how the evacuation was to be carried out. Naively or logically, which amounts to the same thing, he imagined that there would be an airlift both for his men and his equipment. All he would have to do would be to destroy the installations and there was no lack of explosive for that.

What he also did not know, however, was that Carpentier, the commander-in-chief in Indo-China, had his views on that too: 'We will evacuate admittedly, but with honour.'

The two things may seem incompatible. Moreover the decision was to lead to disaster. By refusing an airlift Carpentier was condemning a column of two hundred men to cover on foot fifty miles of road which had long been abandoned to the Vietminh. This was tantamount to sending them into the jaws of the enemy. And so it turned out; the Vietminh was swarming all along the road.

To be successful it was imperative that General Carpentier's evacuation plan remain secret. Naturally it was secret. Equally naturally the Vietminh knew all about it.

Alessandri liked Charton and at the end of the month he paid him a visit. He thought it was his duty both as a commander and a soldier to warn Charton what was afoot. 'They're preparing to play a dirty trick on you,' he said. 'They're going to make you evacuate Cao Bang, but along Route 4.'

'It's madness,' the colonel replied.

He had seen the danger at once and so he sent a telegram to Saigon giving his point of view. 'Don't worry,' came the reply. 'A relief column is being despatched from Langson to join up with you after your first sixteen miles. Once junction with this relief column has been effected, your force will be too strong for the Vietminh to attack you.'

Charlton pulled a long face. He had heard of this 'relief column'; in code it was known as the 'Bayard group'. It was commanded by a tired old colonel nearing the end of his time who, to make matters worse, was an artilleryman. The group consisted of North Africans and they were tired too. Charton knew them well since he had already sent them packing from Cao Bang.

The only reassuring factor was that the group also included a really good unit, the 1st Foreign Legion Parachute Battalion, whose commander, Major Segrétain, was as tough as Bayard was soft. 'Apart from them,' Charton observed, 'the Bayard group is not worth a sausage.'

Further orders arrived from Saigon. 'Evacuate Cao Bang on 3 October. Do not destroy equipment so as not to alert the Vietminh. Move as unobtrusively as possible through the bush on foot. The Bayard group will be awaiting you at Kilometre 28.'

Charton shrugged. He had decided not to obey, first because he feared the Bayard group would fail him and secondly because, if he was to evacuate 'with honour', there was no reason not to do the thing properly, take his guns and vehicles with him and blow up everything behind him. He would go with bugles blowing if necessary; Charton liked a little panache.

On 3 October Charton, the 3rd Battalion of the 3rd Foreign Legion and the rest of the garrison of Cao Bang departed with a flourish. A few minutes later the town blew up.

That same day the Bayard group was already in difficulties. It had left Langson ten days earlier and was due to pass first via That Khe, which was still firmly held, and then halt for several days at Dong Khe before making the final hop to Kilometre 28. On return, reinforced by Charton's column, it would pick up in succession the garrisons of Dong Khe and then That Khe. All would then return to Langson, the final evacuation point.

Unfortunately for the 'Bayards' (and the others) the Vietminh had captured Dong Khe before he arrived. It had been held by two companies of 3rd Foreign Legion and had fallen on 16 September after a two-day siege. On the outward journey, therefore, it was first necessary to retake Dong Khe and hold it long enough to join up with Charton. This operation failed, however, on 2 October. Not only did it fail but suddenly the force was attacked from all sides. The Vietminh army was all around Dong Khe, on watch, hidden in the bush, the quarries, the valleys, the hills.

The country was appalling. Throughout the 3rd the Bayard group fought in small handfuls; it had been split up in the forest into scattered isolated parties. They had tried to take refuge in the jungle but there it was even worse, a nightmare world without tracks, clearings or water points. The radio only carried over 200 yards and one could see nothing. The

Vietminh, ever present but invisible, only showed themselves in order to kill and then disappear. Segrétain's 1st Foreign Legion Parachute Battalion formed the rearguard, beating off attacks and counter-attacking but finally withdrawing, overwhelmed by numbers.

Next day Bayard took refuge in a little valley named Coc Xa. It was in fact a gorge surrounded by precipitous 300 ft cliffs. Here the agony of the Bayard group continued for two days. The Vietminh had infiltrated into the valley, cutting down the Moroccans, splitting up the force, annihilating isolated groups. On top of the cliffs the Foreign Legion battalion, ordered to sacrifice itself to enable the force to continue on its way towards Charton's column, was fighting with its back to the void. Outnumbered by fifty to one Segrétain's parachutists were being pushed inexorably towards the abyss.

'We are asking for authorisation to rejoin the main body,' Segrétain sent to Colonel Lepage commanding the Bayard group.

'Out of the question.'

Segrétain thereupon counter-attacked and held the enemy. His only solution was to attempt a break-through and he was about to do so when a message came: 'Lepage has changed his mind; you are to come down into the "valley".'

Lepage had decided upon this change of plan as a last-ditch measure; he hoped that, once again, the parachute battalion would be able to clear the way for him to emerge from his rocky trap and move out into the bush towards the hills which he could see in the distance and from which Charton should arrive. Unfortunately by the time the order reached the parachutists, night had fallen. They had to come down of course – but at what a price! There was one path running down the cliffs and the Vietminh were already on it.

The sequel is terrifying. The column of legionaries was confined to a mountain path barely three feet wide; on one side was a wall of rock, on the other a void – 300 feet straight down. In the darkness a man would feel a hand; it would be that of an enemy who dragged the soldier with him. Men knifed each other from either side of a bush; they fired at each other guided only by the rifle flashes. The wounded were carried on stretchers, gagged to prevent their groans drawing the Vietminh's attention. But the Vietminh were everywhere; they knifed the wounded as they passed on their stretchers; they hurled weapons down the cliff.

By the morning the parachute battalion had lost one hundred legionaries in this suicidal descent. On top of the cliff there remained only the platoon of NCO trainees. It was too late for them to follow and they were now definitely cut off. So they charged straight ahead. Only seven got through.

Down in the valley the arrival of the parachute battalion had changed nothing. But Colonel Lepage was a believer in miracles. In order to break out of the Coc Xa chasm and rejoin Charton's column he had to carve his way through the encircling Vietminh, moving along a narrow path which was nothing but one long ambush. Segrétain asked that the break-out be made by day so that at least they could have the benefit of air support. Lepage refused; the parachute battalion was to operate by night. It was sheer suicide; there was no hope for a single man to escape. To dive into a narrow passage swarming with enemy in country such as this simply meant that a whole battalion was being coldly condemned to annihilation.

At midnight on the 7th, however, the Foreign Legion Parachute Battalion was in position. Its strength was 210 men out of an initial strength of over 500. The Bayard group took four hours regrouping in order to follow the legionaries into the breach. Dawn was breaking when Segrétain was ordered to attack – too late to profit from any surprise effect, too early for the air force to intervene. The inevitable occurred. A thousand, perhaps two thousand, Vietminh were in position – no one will ever know. This was not an ambush; it was an execution.

Hardly had the legionaries moved off than they were mown down by machine guns enfilading the path. From the flanks invisible rifles fired at them point-blank. Yet they moved on, machine gunning, hurling grenades; they paused and then moved on again – a burst of fire – a grenade – forcing their enemy back as at Terapegui, as at Bir Hakeim. Men fell and dragged themselves up again. The wounded fought from the ground. Some were taken prisoner and clubbed down by the Vietminh. Others committed suicide.

The break-out was not made by the parachute battalion; in a few moments it had lost 120 men. The break-out was made by the Moroccans of the 8th Tirailleurs Marocains. They rushed forward, not as soldiers but as madmen, demented beings, foam on their lips, eyes starting out of their heads, literally mad with fear. They trampled over the legionaries and tore through the curtain of fire, breaking through the Vietminh without even seeing them. Even when wounded they continued to run. Less than 2000 yards away they had seen Charton's column fighting its way forward in good order. They rushed towards it, not to bring it assistance but to destroy it.

To understand the tragic story, we must now return to the evening of 3 October, four days earlier. Charton and his legionaries left Cao Bang in the morning in a well-organised convoy, legionaries out on the flanks holding vital points, screening and reconnoitring the road. Route Coloniale No. 4 had not been destroyed by the Vietminh. In any case there were not very many of them; they were far away round Dong Khe but Charton, though he did not know this, acted as if they were there.

The convoy moved on in a formidable array, guns ready to fire, tanks closed down in case of ambush. Charton had only sixteen miles to cover before joining up with the Bayard group, which would at least give him numbers if it did not produce quality. By the evening of 3 October, having taken all precautions to guard against surprise, Charton had reached Kilometre 17. Next morning he would cover the remaining distance – a mere nothing. His rendezvous was at midday.

During the night, however, telegrams came in. First he heard that Lepage had been unable to recapture Dong Khe. 'The rendezvous is no longer at Kilometre 28 but east of the old post. You will assist Lepage to hold this point.'

At 4 a.m. came a further telegram: 'The Bayard group has been attacked. To avoid annihilation it has moved southwards.'

Charton pulled a face. He knew the south; this was the bush and at the end of it was the Coc Xa gorge. Perhaps Lepage had done the right thing but, Charlton thought, if the word 'annihilation' was being used, the Vietminh must have gone all out.

At 5 a.m. yet a further message: 'Bypass Dong Khe by the south. Junction with Bayard at Quang Liet.'

Charton looked at his map. Quang Liet was a tiny dot south of Coc Xa. Running along the hilltops the Quang Liet track led straight to That Khe, fourteen miles further west. It was perhaps a good idea but it meant that the Legion's column would have to take to the bush.

In addition to his soldiers, legionaries, tanks and Moroccans, Charton had at least 2000 civilians, Vietnamese inhabitants of Cao Bang who had no wish to remain in communist hands. In the morning he destroyed his equipment and the column started down the Quang Liet track on foot. It was an appalling track – in Indo-China vegetation can submerge an aerodrome in three months, so what can it do to a forest track! The legionaries hacked it open again. To give an idea of the length of Charton's column and the speed at which it moved – its head entered the jungle at 7 a.m.; at 3 p.m. a queue of civilians over half a mile long was still waiting to start. Nevertheless the legionaries moved forward at a thousand yards an hour.

Charton followed the Quang Liet track for three days. On 6 October he emerged from the jungle and climbed the hills opposite. There he could hear the firing in Coc Xa about two miles away and over his radio he heard Lepage's calls for help from the bottom of his gorge.

'Faster,' Charton ordered. He had realised that the roles were now reversed; it was now for him to go to the aid of his 'relief column'.

Throughout the 7th he advanced but with greater and greater difficulty. The Vietminh were there. Fighting all the way, the column broke through, manoeuvring impeccably as usual. One after another the key hills were captured. By the morning of the 8th Charton was above Coc Xa. Over the radio he had heard that Bayard was about to try to break out.

Then, about 9 a.m. he saw the Moroccans coming. In their terror they at once spread disorder in the ranks of the North Africans of Charton's column. Nothing is more contagious than panic. In a moment fear had gripped the hitherto orderly ranks. Out in front the legionaries of the 3rd Battalion unflinchingly held off the enemy assaults but in rear was total disorder.

'Who's in command of this rabble?' Charton yelled. No one answered; the officers of 8th Tirailleurs Marocains had torn off their stripes. Lepage arrived. Charton received him coldly: 'Organise your circus,' he told him. 'These are soldiers you are commanding, not sissies.' It was no good. Lepage was in tears, saying: 'And to think that I was to be promoted "substantive" colonel this year.'

At this point Charton had had enough. He realised that all was up. He went off by himself to join his last legionaries still holding out against the enemy waves but isolated out in front. Rifle in hand he fought with his men like any infantryman. He was wounded but carried on. He was wounded again but still did not give way. He was made prisoner that evening after being wounded for the third time; even then, to capture him, the Vietminh had to club him down from behind. With him were three legionaries. All the rest had vanished.

The Foreign Legion Parachute Battalion had vanished too; only some ten survivors managed to gain the sanctuary of That Khe.

March 1954. Ever since 20 November 1953, in other words for three and a half months, the French had been at Dien Bien Phu. Initially the garrison had been the preserve of General Gilles's parachutists but it had been progressively reinforced by infantry brought in by air. Early in December two battalions of the 13th Half-Brigade of the Foreign Legion (the famous 13th DBLE) had arrived. The commander was Colonel Gaucher, a six-foot giant with the figure of a furniture remover and a voice of thunder. He invariably wore a *képi*, refusing, like his forbear Colonel Rollet, to wear either a steel helmet or a beret. He carried a cane, or rather a sort of bludgeon with which he emphasised his words when issuing orders. He was a hard drinker, slept hardly at all, was always on the spot and knew every legionary in his 'shop', as he called it. He was the life and soul of his unit and under his command the 13th DBLE drank hard and swore even louder than most.

After these two battalions two further Foreign Legion battalions arrived: the 3rd Battalion of the 3rd Foreign Legion which had been re-formed after the affair of RC4 four years before, and then its 'Siamese twin', the 2nd Battalion of the 3rd Foreign Legion. These two battalions were commanded by Colonel Lalande. He too was a 'card', perhaps less highly coloured than Papa Gaucher but equally effective. Calm, impassive but brave, he took personal command on that tragic day of 8 May when resistance ended with the legionaries' final charge from the strong-point 'Isabella'.

In December another Foreign Legion battalion arrived in Dien Bien Phu: the 1st Battalion of the 2nd Foreign Legion under Major Clemençon.

From the very first days of the establishment of the post

another Foreign Legion unit had been there, also a survivor of the Dong Khe disaster: the 1st Foreign Parachute Battalion, commanded by Major Guiraud. Early in April, at the height of the battle, another battalion of the Legion arrived, the 2nd Parachute Battalion, eager to win laurels equal to those of the 1st, its senior. It was commanded by Major Liesenfeld but it had only a single month in which to prove itself and show its courage. On the final day the survivors of the two parachute battalions, combined into a company numbering only twenty-two men, launched the final assault on Éliane 4.

One battalion of the Legion, though burning to join the battle, never reached its goal – the 2nd Battalion of the 5th Foreign Legion. On 22 March Colonel Langlais commanding the parachute troops in Dien Bien Phu received the following message through 'unofficial channels': 'My battalion, which is not parachute trained, volunteers to drop *en bloc* into Dien Bien Phu. Stop. Signed Cabaribère, Battalion Commander.'

Langlais smiled. He knew Cabaribère well. Both from his name and his accent he was a southerner, a man of steel, never really happy except when in a tight corner. He had been held prisoner in Laos after a Vietminh ambush and had succeeded in doing what was thought to be impossible – escape across nearly two hundred miles of bush. Cabaribère and his battalion never parachuted into Dien Bien Phu; it was to be annihilated a few days later at Ban Yen Yam on Route Coloniale No. 5 between Haiphong and Hanoi.

The battle of Dien Bien Phu is still vivid in the memory – fifty-seven days of incessant fighting – heroic exploits great and small – days of fire, hunger, death and mud. The real detail of the battle cannot be given here; incidents are too numerous and too similar in the dry language of official reports: 'Attack on strong-point "Huguette 6". Attack commenced 1730 hours. Position surrounded by Vietminh by 0800 hours.

Centre of resistance round command post' – 'Counter-attack at 0400 hours' – 'Enemy repulsed at dawn.' That is all – or nearly all. Figures of casualties are given – terrible figures, a daily average being 'sixty killed, one hundred and twenty wounded'. And what about the other side? No one knew. Perhaps six hundred, perhaps eight. But it was of little importance, for just as many would be there tomorrow or the next day – always provided that one was still alive to see them and fight them.

Fifty-seven days of battle – they cannot be described in a few lines.

Dien Bien Phu, 30 April 1954. The official communiqué on the day's activities mentions only the main occurrence. '1st Battalion of 13 DBLE raided Vietminh works south of Éliane 2. Two blockhouses totally destroyed and one damaged.' The true story, however, is worth telling.

Dawn broke grey and dirty over Dien Bien Phu. The night's fighting had raised a fine black and red dust which fell slowly like volcanic ash and obscured the hilltops. It was still too early for fighting to start again. Each side was slowly recovering and preparing for the next battle. Nothing was to be heard but the occasional crack of a rifle or the dull thud of a solitary shell. Aircraft had not yet left Hanoi, so the sky was quiet.

On Éliane 2 (Hill 5 as the Vietminh called it) was a jumble of enemy trenches, close together and almost at grenade-throwing distance. In this area units were relieved every forty-eight hours, for continuous fighting makes inroads on strength. On this 30 April the strong-point was held by a company of the 1st Battalion, 13th DBLE.

'It's Camerone day, Captain.' Wagner, batman to Captain Coutant commanding this particular company, had entered his CO's 'blockhouse'. It was in fact a miserable shelter, a semi-ruined hut jammed between the roots of some tree or other. Wagner had brought a mug of steaming coffee. How he contrived to make it was one of the mysteries which Coutant had never managed to discover. Whenever asked, Wagner merely gave the enigmatic reply: 'By muddling through, Captain.'

On this particular day Wagner's 'muddling through' had done even better; beside the tin mug was a blue packet of Basto cigarettes – and it was a full one. 'It's Camerone day, Captain.'

Coutant passed a hand over his unshaven chin. Camerone! He had nearly forgotten. For a moment there passed through his mind a picture of Sidi-bel-Abbès, some eight thousand miles away, a picture of the place on holiday – the march-past, the parade, the fête and then the beanfeast. He pulled a face. Some memories are better suppressed when all there is to eat is a plate of rice with a few peas (a 2 lb tin had to do for eleven men) and a sardine. And what for drink? There was not even 'vinogel', the result of a process which dehydrated wine. It produced a peculiar purplish jellified substance with a high alcoholic content which the legionaries would swallow down neat when there was any, for finding any water on Éliane 2 was quite a feat.

'A good Camerone day, Captain.' The platoon commanders had arrived and stood there respectfully according to tradition despite the mud, the battle and the 'circumstances', their excuse for their somewhat dishevelled appearance. Tradition was observed to the letter. Coutant read the account of Camerone, raised his tin mug containing one or two drops of the precious '*vinogel*' specifically kept for the purpose, and sang 'Le Boudin'.

On this day the Legion was ready. If the Vietminh attacked they would find themselves facing legionaries determined to repeat history; none of them reflected that the 13th DBLE had in fact now been reliving Camerone for forty-three days.

In the trenches outside some of the legionaries had their noses in the air gazing at the sky. 'What are you waiting for?' asked a sergeant. 'The aeroplanes. They're sure to drop us some wine and cigars. One has to celebrate 30 April.' The sergeant shrugged – no point in arguing. Everyone knew that in the staff's calculations arms, ammunition and grenades came ahead of red wine – and Éliane 2 used some three thousand grenades a night.

'Captain!' At the door of the shelter, gasping as if he had just run the hundred yards, appeared Fröhlich. He was a broad-shouldered blond giant, his hatchet-face showing hardly a wrinkle above a budding beard, a picture of a legionary fit for a recruiting poster, people said.

'Captain, a package of *"vinogel"* has come down in front of the position, on the Vietminh side.'

'So what? You're not going over there.'

'No, but I just thought it a pity . . . particularly since the Vietminh are like us – nothing to drink.'

Coutant smiled. In view of the circumstances the legionaries found it understandable that they could not drink to Camerone, but they would never agree that the other side should not be in the same boat.

'OK,' Coutant sighed. 'Bring up a machine gun and stop anyone getting near that package.'

So throughout the day of 30 April four legionaries kept watch over a number of packages of bad wine.

The situation report that day registered only minor activity. About 3 p.m. a Chaffee light tank received a direct hit from an enemy 105 mm. There was no harassing fire and no attack – a blank day. It was as if the Vietminh had wanted to present the legionaries with a day off, the only possible relaxation.

In fact the Vietminh were regrouping. General Giap had concentrated his assault troops a few miles from the Muong Thanh plain to indoctrinate them with revolutionary faith.

On Éliane 2 night fell with its accustomed suddenness. The

'*vinogel*' was still there, some fifty yards from the forwards posts – so near and yet so far – a real temptation.

The Vietminh attempted a sortie. It failed. Between them and the legionaries this had become almost a matter of principle. It was in danger of becoming an obsession and Coutant wondered whether, in the words of Soos, his sergeant-major, he ought to 'organise what he could not prevent'. So, after a brief consultation with his platoon commanders, he decided to select from the volunteers (practically the entire company) a team to go and recover the famous packages – 'If we don't go and get them, someone is going to make a BF of himself.'

It was 10 p.m. – a night like most of the others – hardly less calm. A Dakota flew down the landing strip dropping its usual cargo of reinforcements, a company of the 1st Battalion Foreign Legion Parachute Regiment this time. The aircraft was invisible but its track was obvious from the stream of Vietminh ack-ack tracer bullets which followed it. On Huguette 2 at the end of the runway a furious battle was no doubt taking place but on Éliane 2 there was nothing.

Still nothing. The little six-man team took up position. A little silent signal was given – a raised hand. The legionaries dashed forward, dodging from shell-hole to shell-hole, creeping along the barbed wire, taking cover behind tree-stumps, heaps of debris or corpses. Duties had been divided – one group the packages, another to create a diversion.

Fifty yards is not a long way on an ordinary street but it is a very long way in a no-man's-land crammed with booby traps and swept by the fire of hundreds of rifles. Nevertheless the legionaries reached their goal.

A whistle signal. The diversionary section went into action hurling grenades at the slits of the Vietminh blockhouses, demolishing them with explosives found on the bodies of enemy 'human bombs'. Meanwhile, passed from hand to hand, the '*vinogel*' reached the French lines.

30 August 1954. Prisoner convoy No. 42 had been on the march for three days. The armistice had been signed and now, after four months of imprisonment, the men captured at Dien Bien Phu were on their way back to freedom.

These four months had been the most terrible of their lives. The phrase 'green hell' was an apt one; after marching them for 400 miles the Vietminh had set up prison camps in noisome valleys where the climate was murderous and men already weakened by battle, marching and hardship, fell like flies. Eaten by vermin, devoured by mosquitoes and a prey to any form of virus, the prisoners died in hundreds. Of its strength of 400 on departure, Convoy 42 had lost 92 on the march and 244 in the POW camp. On this 30 August 1954 seventy-three survivors were dragging themselves towards the goal for which they had hoped so long – freedom.

From Chieme Hoa above the Claire River they had taken three days to reach the regrouping centre, the last stage before their hand-over to the French authorities at the little port of Vietri. At this moment they had just entered a devastated village where bamboo shelters had taken the place of solid houses. Sometimes the remains of a wall were distinguishable or the flooring of a room or the steps of an old stairway. Apart from ruins, nowhere more than six inches high, nothing remained of what had once been Tuyen Quang. The Vietminh had flattened it themselves; apparently it was a favourite target for air bombing.

Tuyen Quang! The name awoke vague memories in the minds of some of the legionaries in the convoy. 'Tuyen Quang' meant something to them, like a verse from a marching song. One of them, Lance-Sergeant Margot, was humming to himself:

Au Tonkin, la Légion Étrangère
A Tuyen Quang illustra notre drapeau.

In Tonkin the Foreign Legion
Will make our flag famous at Tuyen Quang.

The legionaries of Camp 42 habitually gathered around Corporal Plewa, a resilient Pole. There was Margot, a Swiss whose white hair and beard had won him numerous favours from the Vietminh (though he was actually only thirty-five, they thought that he was a decrepit old man with one foot in the grave); then there was Gimber-Rus, a Spanish lad who had been in the heavy mortar platoon of the Foreign Legion Parachute Battalion and was short of an arm; then Ennen, a skin-and-bone German who had lost two and a half stone in four months; finally there was Ballestracci, known as 'Mimile', a young sergeant who had parachuted into Dien Bien Phu on one of the last days.

'Tuyen Quang. What's that anyway?'

Margot rummaged in his memory and then told the story. 'It happened at the end of 1884. The French had landed in Tonkin and, having signed a peace treaty with the Chinese, were busy cleaning up the area, especially this one, from the freelance bandits who were terrorising the population. As we did in this war, units were established in posts from which they worked outwards; you know the game – patrols, reconnaissances, opening up tracks, beginning to build roads.'

'Not much has changed,' Plewa observed.

'Nothing in fact. Moreover the Legion was there. It had taken part in the assaults on the fortresses of Sontay and then of Bac-ninh. Its enemies were the Black Flags, Chinese bandits, half brigands, half soldiers.'

'Why "Black Flags"?'

'Because they always went into battle behind enormous black silk flags. Captain de Borelli captured two of these flags at Bac-ninh and presented them to the museum in Sidi-bel-Abbès.'

'And what of Tuyen Quang?'

'De Borelli was there too. Early in 1885 the Chinese began

concentrating round the post which was on a hillock overlooking a little plain regularly flooded by the Claire River, as it still is today.'

Commanding the legionaries in the post was a brave energetic officer, Major Dominé. His total strength was two companies of the Legion, one company of Tonkinese sharpshooters and eight sappers under Sergeant Bobillot.

The Chinese surrounded the little fortress with a line of trenches based on fortified villages. On 26 January they bombarded and set fire to Tuyen Quang, then launching three columns against the blockhouse defended by Sergeant Leber. They were driven back everywhere but their investing line was now only five hundred yards away. On 30 January the blockhouse was evacuated.

From this point matters developed quickly. The enemy gained ground rapidly and was almost at the walls of the fort. Loopholes became untenable and had to be protected by wooden palisades which the Chinese tried to tear down with grappling-hooks. Attacks became increasingly frequent and the bombardment more intense. Ammunition was running low and firing was only allowed if a hit was certain. The Chinese earthworks were so close that sentries could not look over the top of the wall without being in mortal danger. The legionaries were forced to revert to the old-time machicolation method.

By 10 February the Chinese saps were so close that a breach in the wall was expected at any moment. By the 11th legionaries and Chinese were literally face to face. Two days later the enemy blew up the south-eastern corner of the fort.

Legionaries and the sappers hastily plugged the gap but on the south side the Chinese had laid charges under the walls intending to ignite them simultaneously and so bring down the entire perimeter. Major Dominé, however, had already ordered a second perimeter wall to be constructed sixty yards to the rear.

The time of decision drew near. On the 17th Captain Dia,

commanding the Tonkinese, was killed and on the 18th Sergeant Bobillot. Four days later the perimeter wall blew up and the Chinese rushed in. They were met by Captain Moulinay's company; a mine exploded killing twelve legionaries and wounding twenty-six but the attack was repulsed.

On the 24th came a further attack, a terrific affair on a pitch-black night. The entire Chinese army of some 10,000 men was involved. Facing them were the last reserves under Captains Cattelin and de Borelli. To the sound of the bugle they charged with the bayonet. At 4 a.m. the enemy withdrew.

On the 25th another mine exploded. The perimeter wall was now no better than a ruin. Yet another attack ended in failure.

On the 28th the Chinese let loose their 'big bang' – over 200 lb of explosive. Pieces of the wall sailed a hundred feet into the air. For four hours men shot point-blank at each other or used cold steel. In place of grenades the Chinese threw sacks of burning powder. It was no use. The attack was thrown back once more. The attackers lost ground as they abandoned the breaches piled high with their dead.

This was the last assault. A relief column under General Giovaninelli was already on its way from Yuoc six miles to the south. Led by its commander the garrison of Tuyen Quang sallied out, brushed the last 'Black Flags' aside and went on to meet it. On 3 March the survivors of the garrison presented arms before the ruined walls of their fort to their saviour, General Giovaninelli, and his 2000 men.

For nearly two months 500 French soldiers had held out against 10,000 Chinese.

'So you see,' Margot ended, '1885–1954 – the Legion has finished its campaign at Tuyen Quang. We've come full circle. This is where it began.'

'Lumme,' Gimber-Rus said admiringly. 'They held out

from December to March. Almost as long as we did. But they had better luck; they were relieved in time.'

'Actually,' Plewa asked, 'isn't this the memorial put up in honour of the men who fought at Tuyen Quang?' He pointed to the remains of a pillar surrounded by a bamboo fence.

'Perhaps.' The five men went up to the fence and saw that it was in fact a memorial put up in 1927 by Major Maire in memory of the garrison and one of its heroes, Sergeant Bobillot.

'The swine!' Gimber-Rus shouted. Whether from a sense of hygiene or a desire to desecrate a 'colonialist' memorial the Vietminh had chosen to instal their latrines at the foot of it.

So on that 30 August 1954 the legionaries of Convoy 42 took turns to stand strict guard over this memorial. They forbade anyone to approach – or use it – despite all the Vietminh threats.

The war in Indo-China was over. One by one the battalions of the Legion returned to Algeria. Another war was beginning there. Oddly enough, however, the final incident of the campaign took place far away – before a court martial in Tel Aviv.

In May 1958 the Israeli naval court martial found before it a young man aged twenty-six in sailor's white uniform. He was Eliahu Itzkowitz and was being tried for desertion.

His was an exceptional, in fact an embarrassing, case. He had jumped his ship in peacetime in order to enlist in a fighting unit and serve in the Foreign Legion for five years including twenty-seven months in Indo-China.

In 1939, twenty years earlier, Eliahu Itzkowitz had been leading the carefree life of a small boy in the little Romanian town of Chisinau. So it might have continued, had not his country's leaders decided to cast in their lot with Nazi Germany at the beginning of World War II. Antonescu, the 'Conducator', took pride in imitating his allies and masters and accordingly surrounded himself with an 'Iron Guard' clad in brown shirts, whose methods soon rivalled those of the Gestapo. Such was their zeal, in fact, that they soon succeeded in shocking even the Germans, not so much by the magnitude of their massacres as by their 'negligence' about burying the bodies. The RSHA (Reichssicherheitshauptamt, the administrative headquarters of the Nazi police which was responsible for mass executions) lamented one day that the Romanians 'did not keep up to date the registers of executions carried out and indulged in unbridled looting'!

The Itzkowitz family suffered the fate reserved for Jews. They were taken to a concentration camp, perhaps not as modern as Auschwitz but equally murderous. There they were handed over to certain of their fellow-citizens who specialised in mass murder and all of them vanished quickly. The man responsible was named Stanescu.

There was one survivor, now head of the family and bearer of the name. When liberated in April 1945 he was a fifteen-year-old skeleton – Eliahu Itzkowitz. He had survived and he had forgotten nothing. He had seen his family die and he knew the name of their murderer; he had sworn to avenge them. In addition to the simple instinct of self-preservation it was the fury of his hatred which had carried him through hell.

As soon as he was free Eliahu set off in pursuit, from one town to another. Obviously Stanescu – the name mattered little since he changed it all the time – had not returned to his native town. He had fled, but whither?

One thing was certain, he was still alive. Itzkowitz followed his tracks pertinaciously, gaining on him slowly. One day he thought that he had succeeded. Stanescu was reported in a town in western Romania. Eliahu went there but arrived too late and found only the murderer's son, whom he stabbed with a butcher's knife. He was arrested and brought to court in Bucharest in 1947 and was sentenced to five years in a reform school. When he emerged in 1952 he was aged twenty-three.

When permitted to emigrate to Israel Eliahu Itzkowitz was immediately called up. He enlisted in the parachute troops and was sent to a training camp south of Rehovoth. He was captivated by his new life; in the desert sun his desire for vengeance faded; his previous life became a vague memory, as distant as Romania, his native country.

But fate did not let him go. When on leave Eliahu met émigrés from every corner of Europe, including some Romanians. They exchanged news of their country and of their friends; they talked of the concentration camps too, the memory of which came back vividly as soon as the subject was mentioned. One day some new arrival let fall the name Stanescu. 'The bastard succeeded in getting away before the Russians arrived. He got across to West Germany, trying to pass himself off as a refugee. Someone recognised him, however, and we were alerted – there are many of us after him. Unfortunately he did not wait for us; when we got there he had vanished again.'

For Eliahu the name Stanescu was like a red rag to a bull. He was now no longer the young immigrant in Israeli parachutist's uniform. He was on the trail again.

A few days later he got information from another Romanian: 'Stanescu. He was lucky. He reached Offenburg in the French zone of Germany and enlisted in the Foreign Legion. The French are fighting in Indo-China, you know.'

By the next day Eliahu had made up his mind. He asked to see his officer and requested a transfer to the navy. Shortly afterwards the transfer was agreed and Eliahu boarded an Israeli Navy destroyer at Haifa. Some further weeks went by and his ship left for Italy to take delivery of certain military equipment. Then matters developed quickly. At Genoa Eliahu requested and was given permission to go ashore. He bought a coat, took a train and crossed the frontier without difficulty. The next day he was in Mentone and the same evening in Marseille.

Less than a week later the *Sidi Ferruch* took to Algeria its normal consignment of recruits for the Legion. Among them was Eliahu Itzkowitz.

During the crisis period of summer 1953 the legionaries' training period in Saida had been cut short. By September Itzkowitz was on board the *Pasteur* heading for Indo-China.

Once there he had no difficulty in finding Stanescu's trail once more. The Legion included only three main infantry formations and they contained very few Romanians. By the beginning of 1954 Eliahu felt that he was nearing his goal; he had located his enemy in the 3rd Foreign Legion. Stanescu had changed his name but Eliahu was sure that he was his man.

A request for a transfer and one or two astute manoeuvres finally brought Itzkowitz together with the man whom he had hunted incessantly for ten years. Oddly enough, he now had no feeling of hatred. Of course this was the same Stanescu who had previously been quite ready to murder so many of his fellow-countrymen but Eliahu felt subconsciously that something had changed. Physically Stanescu was about the same. Active service had ensured that his figure had not changed. Morally, however, he seemed to be of a new mentality. He was now a corporal in the Foreign Legion, entirely unrelated to the executioner of Chisinau.

Nevertheless Itzkowitz did not hesitate. He had sworn vengeance and the execution would take place – he felt that it was his responsibility. Eliahu, moreover, was a good soldier. Tough, resilient and brave, he was always ready to volunteer for a patrol. He was looking for his opportunity.

It came in June 1954 on Route 18 between Bac-ninh and Seven Pagodas during one of the Legion's last battles in Tonkin shortly before the cease-fire. The company had been ordered to open and hold the road long enough for a supply convoy to get through. It was a routine job but a murderous one, for hardly a day passed without the Vietminh attacking, sometimes in force.

On this particular day Eliahu was the leading scout and he pushed his way through the reeds, his gun at the ready. He knew that his corporal was immediately behind him. At that moment the Vietminh opened fire.

Stanescu dashed forward, knelt beside his scout and observed the countryside. Itzkowitz looked round to ensure that they were alone – 'Stanescu?'

The corporal hesitated for a moment – he had not heard this name used for a long time. Slowly he turned, his eyes showing surprise rather than fear. 'You really are Stanescu, aren't you?'

The corporal did not understand but he nodded. 'Yes, but . . .'

'Stanescu,' Eliahu went on in an icy voice, 'I am one of the Jews of Chisinau. I have come here to serve the memory of your victims' – and with that he emptied the magazine of his tommy gun.

A burst of fire came from the Vietminh and, at the risk of his life, Eliahu dragged the body back to the road. A legionary will never abandon the body of a comrade.

Eliahu Itzkowitz had kept his oath. He finished his tour of duty and served out his five years with distinction. Then he returned to Israel. On arrival at Haifa he was met by two military policemen who took him straight to the military prison.

The Israeli authorities at first had great difficulty in believing his story which they considered fanciful. When consulted, however, the French authorities confirmed it.

The Tel Aviv court martial was accordingly lenient. Eliahu Itzkowitz was sentenced to only one year's imprisonment.

Benir Smir

January 1960

'REGIMENT ! ATTENTION !'

The order rang out as crisp as the man who gave it. Colonel Gardy was no colossus but a spare man, his face lined by a long career in the Legion, a sharp beaky nose between bony cheeks. Around him the legionaries of the 1st Foreign Legion were drawn up in square in the Viénot barracks in Sidi-bel-Abbès.

In a voice dulled by emotion Colonel Gardy addressed his men, both young recruits and old hands waiting for release. 'Dien Bien Phu has just been taken. We are assembled here to do homage to the sacrifice of those who have fallen in this epic struggle. We shall now present arms to the colours which have vanished in the battle.'

A silence and then the colonel called out as if rendering a report:

'The 2nd and 3rd Battalions of the 13th DBLE.
The 2nd and 3rd Battalions of the 3rd Foreign Legion.
The 1st Battalion of the 2nd Foreign Legion.
The 1st and 2nd Battalions of the Foreign Legion Parachute Regiment.'

A further silence and the 'Last Post' rang out. The Legion was united in spirit at this moment. There was not a legionary present who would not have wished to be with his comrades.

The war in Indo-China had just ended and for a time, until the Vietminh had handed back the last prisoner, the Legion consisted only of rear bases – and strength tables.

The months went by and the armistice was signed. Orders arrived from Paris to the effect that the Legion had no need to retain so many soldiers. It was to be reduced from 30,000 to its peacetime strength of 12,000 legionaries. In Marseille young recruits found it difficult to enlist – only one out of four volunteers was accepted – and in Sidi-bel-Abbès re-engagements were discouraged.

By the end of 1954 the first prisoners returning from Indo-China were arriving in Sidi-bel-Abbès. They caused anxiety and sometimes fear. Their period with the Vietminh had changed them; they were broken in spirit and marked in body. They were suffering from a sort of sickness of the mind, the 'yellows', and its symptoms were nostalgia for the country they had lost, ill-will towards their leaders and mistrust of anything which did not belong to their little world. They kept themselves to themselves, took their decisions in concert, only went out with each other and answered questions in unison. They had acquired the habit of reflection, of self-criticism – sometimes just of criticism. In short they were suspect, in some cases ungovernable.

Towards the middle of 1955 formed units began coming back in their turn. The headquarters of the Legion looked at these prodigal sons and could barely recognise them. The imprint of Sidi-bel-Abbès was no more than a thin veneer. The Indo-China campaigns had changed the spirit of these units. They had acquired ideas of independence and habits involuntarily copied from the enemy. In Indo-China officers and NCOs had lived a bachelor life, sharing everything with their legionaries, living with them twenty-four hours out of twenty-four. They were no longer inspired by an *esprit de corps* but by the spirit of a monastery. They formed a world apart, closed to everyone else. It was imperative to give them back the 'imprint of the Legion'.

The Legion took appropriate action. In 1955 and 1956,

whether at Oran, Bizerta, Tunis or Algiers, the ships bringing the regiments back found the band of the 1st Foreign Legion on the quayside. These men were retaught the meaning of the green tie and the value of the epaulette. They came home to Sidi-bel-Abbès and were re-formed by the inclusion of an admixture of young recruits and very old soldiers. Their officers were promoted and so replaced; other officers less scarred by Indo-China were given the task of reinjecting the spirit of Africa into the 'Indo-Chinese'.

Matters were moving quickly, however, and the regiments which had left Haiphong or Saigon with, in their hearts, a fear of the forthcoming period of peace, found on arrival in Algeria the same climate of insecurity to remind them of the country which they had just left. There were the same street patrols, the same barbed-wire barriers at crossroads, the same anti-grenade screens in front of the cafés. The Algerian revolt had begun.

'We know this game,' said the old soldiers with a shrug of the shoulders, 'but it won't be the same here; the Arabs are not the men the Vietminh were.' Nevertheless the war was to last seven years.

Units of the Legion had to make their contribution straight away. After a brief spell in Sidi-bel-Abbès, time enough to acquire a different mentality, they were sent to the hot spots of Algeria. Immediately the legionaries realised that this emergent war bore only a distant relationship to war as they had waged it hitherto.

They perceived that the 'rebellion' – the official term used to conceal the fact that it was open warfare – had become a cancer in the country. The 'miserable wogs', yesterday no more than a joke, had taken to planting bombs and cutting the throats of women and children. These lazybones proved highly active in sawing down orange trees and tearing up vines. These cowards who had fled from the Vietminh sometimes proved to be formidable fighters. War had changed both people and things. Though only mediocre as mercenaries, the Arabs suddenly showed spirit when fighting for the freedom of their own country.

At the day-to-day level maintenance of order degenerated into a monotonous series of small chores. It was lonely too. Algeria is a pleasant country only in summer – even though it does last six months – and along the seashore. The hinterland consists of mountains covered in thorn-bushes, waterless pebbly djebels, muddy roads, ugly filthy Arab villages, and Europeanised towns, joyless, unimaginative and unattractive. Compared to the Vietnamese villages which were fresh and welcoming with their open houses and their clean polite inhabitants, the legionaries found these enclosed *mechtas* sordid and lousy; with their lack of windows and dubious smells they concealed a repellent swarm of humanity. Indo-China had been a sister country; Algeria was a foreign province.

The legionaries sighed but went off to war again, sometimes with their hearts in their mouths. They were left little time to ruminate, however. As soon as they were operational they were despatched to the critical areas alongside their colonial parachutist counterparts. From the Aurès mountains with their steep Colorado-like valleys to the bare and forbidding Nementchas, from Ouarsenis to the Sahara they held the hills, opened the tracks and combed wadis and *mechtas* in their search for the elusive *fellagha*. Round a hedge corner a casual shepherd would loose off a charge of buckshot from his sawn-off shot-gun at the leading scout; the legionary would be cut to pieces; the shepherd meanwhile vanished. The search for the enemy was a sordid desperate business.

Between July and December 1955 the Foreign Legion assumed its role of wartime 'fire brigade'. It took part in major operations with curious code names – 'Whistle' south of Palestro, 'Hammer' and 'Hurricane' in the wooded massif of the Collo peninsula and 'Tango' along the valleys of the Aurès mountains. It learnt to its cost that any colonel seeing a Foreign Legion regiment arrive in his sector invariably took advantage of it, allotting it tasks which he would not dare give to his own troops. For him it was an understood thing: the Foreign Legion had just arrived and they were 'fresh troops'. It did not matter to him that before its arrival the unit in question had been on the march for weeks at the other end of the country, that its uniforms were in rags, its boots worn out and its men at the end of their tether. Off they went again.

The Legion began to feel that it was always pulling other people's chestnuts out of the fire, these young conscripts who lorded it over the villages and filled all the recreation areas. In Algeria the Legion was invariably quartered outside the barbed wire surrounding a town; as soon as it arrived the Arabs shut up shop and the European cafés raised their prices for beer and other drinks. The Foreign Legion was reduced to carrying with it its bar, its cinema, its brothel and its mess.

These are perhaps prosaic details but they had their importance. The Legion did not feel at home in Algeria. With its dusty yellow towns, deserted cafés, unkempt soldiers milling round in the midst of hostile 'wogs' and barbed-wire barriers blocking roads pot-holed by the 'regular army's' great American lorries, it bore no relation to the country of the Legion's early days. The legionaries thought nostalgically of Indo-China where they were 'among their own people'.

In fact, apart from the ports of Algiers, Oran, Philippeville or Bône, the towns of Algeria are primarily designed to discourage the tourist – or the soldier. Anyone who has seen Batna in winter, for instance, will know that this so-called city is a disaster.

Early in 1956 major operations were halted, or rather localised, and the country divided into sectors. The 3rd Foreign Legion, commanded by Colonel Thomas, was in the Aurès mountains. Once more the Legion was busy constructing posts. In contrast to those in Indo-China which were made to protect a village, provide a base for contact with the population and a jumping-off point for attack, the post in Algeria was planned merely as a transit base between two operations. There was no question of attack, still less of any beginning of pacification. Accordingly the majority of these posts were simply an enclosure surrounded by a dry stone wall including – or perhaps not – an observation tower. In one corner was the vehicle park and at the other end the platoon tents. Once the operation was over the unit packed up and moved off to start the same game elsewhere. Naturally the points chosen were the highest – and therefore away from the water points – and those with the best view – therefore open to the wind. So the legionaries were thirsty by day and cold by night.

Sometimes, however, there were surprises, for the Algerian rebellion was no simple affair. One morning in June 1956 the sentry on the post of T'kout shouted: 'Lieutenant, there are some *fellagha* approaching.'

'Open fire,' the officer replied calmly.

He was not afraid of an attack, only of the wind from the sands. T'kout was impregnable; located on top of a rocky ridge, it overlooked the valley of Baniane which was as deep and as deserted as a canyon.

'It's that bastard Veraërt who's drunk too much again,' Sergeant Belling chuckled. 'He's seeing things.'

Nevertheless Veraërt was right. Four men in khaki uniform, wearing balaclava helmets and carrying a white flag, presented themselves at the barrier of the post. On the lapels of their tunics were green and white badges carrying a crescent and red star. They were 'Fels' – genuine ones.

'Well?' the lieutenant asked gruffly.

'Well,' the FLN officer explained – he seemed quite at ease and had settled down as if for an ordinary conversation – 'we have come to ask you to cease all operations for four days starting on Monday.'

The Foreign Legion officer was thunderstruck. Wide-eyed he asked. 'There's no reason to. Why should we?'

The Arab cleared his throat and explained. 'Well now, we have decided to mount an operation in this area to liquidate the rebels.'

'The what?'

'The rebels. You know, these "chaouia" autonomists. They're not even Arabs or Moslems and they killed some of our men a fortnight ago.'

Flabbergasted the lieutenant realised for the first time that there were rebels within the rebels and that they were deadly enemies.

'Have I your word?'

The Legion officer evaded the issue. 'It's not for me to decide. I must report.'

'I'll wait.'

The lieutenant sent a message. Two hours later came the reply: 'Agree to let the FLN act. But afterwards they must come over to us.'

'You're joking,' the FLN officer countered. 'We're not here to do your job. Follow us up if that amuses you.'

He departed. And on the Monday morning a relaxed set of legionaries from T'kout watched the battle as spectators.

The first two legionaries to die for France in Algeria came from the 3rd Foreign Legion. They were named Kreuze and Zelli and were killed on 18 March 1955.

The 13th DBLE was reconstituted, partially with survivors from Dien Bien Phu and partially with new recruits; it was sent to the Tunisian frontier in the Nementcha massif, under the command of Colonel Rossi. The 13th DBLE had always been looked upon as a mountain unit; it shared the area with the Chasseurs Alpins.

The 5th Foreign Legion was spread along the Moroccan frontier while the Foreign Legion Cavalry Regiments were formed into 'scouring' units and patrolled, first along the tracks bordering the frontier and then along the frontier wire.

The situation in Algeria demanded numbers. After the regular units from France came formations of conscripts. By 1956 there were 500,000 men in Algeria. Accordingly the Legion once more opened its recruiting doors wide – but in vain; enlistments had never been so low. The war in Algeria attracted no one. 'There's no end to it,' the old hands said. 'It's boring,' the young ones maintained. They had thought to find in Africa palm trees, sand and sunny excursions; what they got was wind, rain, muddy tracks and ambushes.

Another reason (low be it spoken) was that feminine smiles were noticeably lacking in Algeria. In Indo-China there had been no problem: 'I like you' – 'You like me' – all very simple.

In Algeria the local women were veiled and jealously guarded by a savage husband, brother or father-in-law. The European girl preferred the company of a regular NCO (he was perhaps an engineer and in any case came from France) or alternatively that of a security policeman (he at least earned good danger money) to that of the legionary who was merely a soldier and would receive only a ludicrously low 'death grant' in place of a pension.

The recruiting figures speak for themselves: in 1955 the Legion only accepted 2981 applicants as against 5447 in 1954 and 6327 in 1953. Even more serious was the fact that re-engagements were rare; between September 1955 and July 1956 not a single legionary of 13th DBLE re-engaged.

The reasons were manifold. The men disliked the country; there was no pay increment, reducing pay to the ridiculous level of 190 francs a month; there were no medals (the 'Cross for Military Valour', the outward and visible sign of a mention in despatches, was not instituted until 1957);

finally and most important, the slowness of promotion was not calculated to encourage re-engagement.

Moreover conditions were deplorable. Units were continually on the move and had no real fixed base; Sidi-bel-Abbès was far away and could not absorb more men than it had already. Regiments were therefore forced to lodge with other units – always on a temporary basis. Barracks were consequently always overcrowded. In Bône, for instance, in September 1955 the Tirailleurs barracks, designed for a battalion of 600 men, was housing simultaneously a battalion of the Legion, half a regiment of parachutists and two companies of riot police – 3000 men in all. The barrack rooms looked like the holds of émigré ships, the mess was full to bursting, the kitchens inaccessible, the food horrible and everything else on a similar standard.

The only two units reasonably well provided for were the two parachute regiments. The 1st set itself up in the pine-woods at Zeralda, on the sea some twenty miles from Algiers; the 2nd established itself a few miles from Philippeville, also by the sea. By contrast with the peripatetic life of other Foreign Legion units that of the 'green berets' was the object of envy. Admittedly they always had the hardest fighting; admittedly their losses in the field were higher (the 1st Foreign Legion Parachute Regiment lost one-third of its strength of 1200 during three months' fighting). But they were regarded as rich; in addition to 'air pay', a bonus issued to airborne troops, they drew a 'maintenance of order bonus' which was only handed out grudgingly to other units. In 1956, moreover, between August and December the Foreign Legion parachutists were given a trip to Cyprus and then to Egypt – to enable them to fight . . . in Algiers. During all this time the Foreign Legion Infantry regiments remained tied to their muddy tracks, wind-swept wadis or snowy djebels.

Winter in Algeria has nothing to recommend it – rain, frost and snow above 3000 feet and many of the hills rise to 6000. An ex-soldier of 13 DBLE tells the story of his climb up

Chelia, a 7000-ft peak in the Aurès, on the night of 1 January:

> We left at nightfall. According to sector 'intelligence' the Fellagha were holding an important meeting there. You're telling me! As if the Fellagha were stupider than anyone else! But the intelligence officers did not stop to wonder whether there might not be something incredible about their information. As if the rebel colonels would choose to hold counsel on top of a mountain on 1 January! We crawled up all night with snow up to our thighs. At 1 a.m. the wind got up – and that was the last straw. Some men in my company had frost-bitten fingers; the thermometer must have been down to about –15°C.
>
> Naturally when we reached the top about 3 a.m. there was no one there. The wind was so strong that not even an auk would have stayed up there.

The days selected for operations were usually Sundays or holidays. The reasoning was simple. 'From the point of view of the Fellagha,' the sector commanders said, 'which is the European rest-day? Sunday. So they will not be expecting us. Sunday is therefore the day on which we have the best chance of catching them.'

In theory it was a brilliant idea but in practice it seldom worked out. Although, during the war in Algeria many operations began on a Sunday (or a Holy Day or 11 November), few owed their success to this fact.

Gradually the Legion got organised. By continually banging the gong the regiments finally obtained a base where at last the legionaries could feel at home. For two years, like tortoises or nomads, they had been carrying around with them all their meagre baggage including their walking-out dress and their personal possessions. The ability to leave these things in a rear base helped to give them a stake in the country.

The 13th DBLE, for instance, taking its cue from the 1st Parachute Regiment, made its camp near Bougie a real model of its type. Morale rose sharply: the legionaries now had a goal when out on operations. They had their camp and they began to make plans.

Operations remained the same, however – patrols, ambushes, opening up tracks, occasional man-hunts but seldom any real fighting, particularly not after 1958. The memory of Indo-China began to fade. The conclusion of any operation was hailed as a release. The legionaries returned to their camp. They tidied it, they swept it, they planted flowers, they decorated. They felt happy there.

For the first time since Indo-China, moreover, the Legion could return to its traditions. A legionary loves his folk-lore but he likes proper conditions in which to exercise it; sounding the call 'Le Boudin' on a bare windy hill between two jerricans doing duty for sentry-boxes is, after all, less attractive than doing so before a freshly painted sentry-box with the guard in parade uniform – *képi*, white gloves, belt and ceremonial epaulettes.

The call 'Le Caid' came into its own again. This was a call peculiar to Foreign Legion units; it was sounded only once a day when the colours were within the camp and the only recipient of it was the unit commander – the 'Caid'. Every morning, as he arrived, the full guard paraded at the entrance to the camp. The unchanging ceremony required each legionary to shout his name as he presented arms. The commander saluted and passed slowly down the ranks. Meanwhile, on hearing 'Le Caid', invariably followed by 'Le Boudin', the whole camp stood strictly to attention. The day had now begun.

For officers, to be received with 'Le Caid' was quite an event. A number of conditions had to be fulfilled, the most important being that the titular commander be absent. Then the next most senior officer took over and he would be the recipient of the call. If this was the first time that such an honour had been paid to him, he had to treat the entire mess to red wine at lunch. It was an expensive distinction.

It was during these periods in camp that *esprit de corps* was formed. The midday meal, with all present, was not looked upon as a moment of relaxation but as part of the duty of the day. All officers had to attend. The ceremonial was complex, rigid as a form of service and adhered to like a liturgy. The formality was impressive; certain songs were obligatory – 'Le Boudin' before the reading of the menu, for instance; equally certain words were banned, as were conversations about 'shop'. The junior officers were sticklers for tradition and it was for the subalterns' union – with the senior subaltern at their head – to set the tone. It was of course an impertinent one.

Officers' mess meetings were equally formal and always included singing – the Legion sings a lot. At these meetings songs were even composed which turned into regimental marching songs. The 3rd Foreign Legion had its song, as did the 1st Foreign Legion Cavalry. The parachute regiments had their own and, after the disbandment of the 1st Foreign Legion Parachute Regiment, the regimental marching song was invariably sung standing strictly to attention.

Meanwhile the war went on. Indifferent to the various political upheavals, the Legion combed the djebels and took part in major operations, the most important of which was named *Jumelles* [field-glasses]. From July 1959 the Foreign Legion regiments were everywhere throughout the Kabyle for a period of five months. The winter went by with units taking turns in the various sectors but the 3rd Foreign Legion and 13 DBLE were always there.

In January 1960 the rising known as that of 'The Barricades' broke out in Algiers. For the Legion this was an affair between Frenchmen of Algeria and Frenchmen of France. The only unit involved was the 1st Parachute Regiment, which knew the city.

Hailed by the French of Algeria and ogled by the girls of Bab-el-Oued, the parachutist legionaries listened, became involved and sometimes allowed themselves to be convinced. They loved this bustling, vivacious, warm-hearted capital with its white port facing the blue sea and its streets of which they were masters. For them this was Algeria. So, when they returned to Operation Jumelles in February 1960, they did not forget.

This is the reason why, in the following year, they were among the first to respond to the call of the 'rebel' generals. In the case of these simple legionaries their crime of indiscipline was committed out of loyalty first to their officers and then to their *pied-noirs* [Algerian-born French] friends. Their officers were men whom they had followed for months and would follow anywhere. Some of these officers left for France and then returned, in some cases from Clermont-Ferrand, to take command of their companies again. The legionaries respected them, were sorry to see them go. Under these officers the legionaries launched into an adventure.

For an adventure it was, involving some risk, a great deal of bluff, secrecy and suspense. In a sense it was a great game which had its light-hearted side.

During the night, as the lorries were rolling towards Algiers, platoon commanders – or rather those who had been assembled in time – announced: 'We are going to capture the Government General.' (The regiment had been off duty and most of the NCOs were at home or, like their colonel, on leave.) Only a few officers and NCOs were present.

With all lights extinguished the convoy slipped into the city and stopped before the great iron gates. General Saint-Hillier, commanding 10th Parachute Division, had only been woken up and warned a few minutes before; he was taking matters very seriously and wished to stop anything irrevocable happening. 'Halt!' he shouted. 'Don't start something you may not be able to stop.'

The legionaries moved on headed by their officers. The

general spread his arms in a ludicrous despairing gesture. 'You will have to pass over my dead body,' he exclaimed dramatically, no doubt hoping to impress everyone with the seriousness of the situation.

'We'd rather go round,' a legionary replied – and everyone laughed.

There was much laughter in the early days of what came to be called the 'putsch' of 22 April. This shows that during these four crazy days the Foreign Legion parachutists – and perhaps others as well – had not really grasped the gravity of their revolt. The other side understood the tragedy of the situation at once. Nothing was said but hatred grew, a passionate deep-rooted hatred, all the more irremediable because certain people had shown cowardice and fear.

Between the two camps the gulf was deep. The legionaries became arrogant, thinking themselves the victors. They swaggered about as if the city belonged to them. They did not know that in the djebels, on the tracks or in their posts, their fellow-legionaries who were less involved and more clear-headed, were listening to the news with stupefaction. 'What are they getting mixed up in?' they asked.

The answer came from Sidi-bel-Abbès, from Colonel Brothier, commanding the Foreign Legion, and he gave it with a heavy heart to little General Gardy, his former commanding officer who had gone over to the revolt: 'Being by definition foreign, the Legion will not intervene in a purely French affair.' These few words summed the matter up.

In Algiers morale began to fall. The 2nd Foreign Legion Parachute Regiment had arrived from Philippeville to support its 'elder brother', the 1st, but it left under cover of darkness to return to its garrison – and its colonel whom it had 'forgotten'. Meanwhile the 1st Parachute Regiment, after considering 'doing a Camerone' on the heights of Algiers, took the road back to Zeralda.

The sequel is well known. Certain officers did not give up and took to 'the maquis' with a few legionaries. The majority of those missing, however, were simple soldiers who had been terrorised and exploited and were now lost – in the literal sense of the word.

From Sidi-bel-Abbès Colonel Brothier hastily despatched to Algiers officers of the Legion to recover these 'deserters'. In the majority of cases the point of no-return came when they were either arrested by the gendarmerie or absorbed by some activist clandestine network. Colonel Brothier's object was to find them, reassure them and then escort them to Sidi-bel-Abbès. The Legion is insistent on settling its own affairs.

Captain Pillet, for instance, in a race against time with the authorities, the police, the gendarmes and 'loyalist' troops, succeeded in filching from Colonel Lavergne, commanding the Blida sector, a dozen legionaries who had been incarcerated in the civil prison and were about to be transferred to Algiers. At the same time the 1st Parachute Regiment left Zeralda to a final ovation from the local populace. They had played a game and the game was finished. The officers were court martialled for mutiny; legionaries were not punished; they were 'amalgamated', returning to anonymity in infantry regiments.

The war was nearing its end. Whether right or wrong the solution imposed from Paris was inexorable. The legionaries obeyed once more, even if they did so with a heavy heart.

'I had a choice,' an officer said, 'and I stayed with my company. If I had "gone across", in no circumstances would I have taken any of my legionaries with me, not even my batman. My feelings are no concern of his.'

Many thought this way. They could not accept a *fait accompli* and they refused to comply with it. So they resigned. There was no shame in resigning once the war was over.

When a Foreign Legion officer leaves his unit, the regiment does a great deal – and in style. Tradition demands that he should spurn the ceremony, and that his comrades see him onto the boat. So his men parade at the port, in full dress with the regimental band and with all NCOs present. The departing officer appears and there, in the midst of the usual throng associated with the sailing of a boat, is the Legion as if on its own barrack square. For the moment the quayside has been taken over by white *képis.*

The band plays and the senior subaltern escorts his fellow-officer – the supposed defaulter – to the foot of the gangway. There his batman is waiting and he has arranged everything as usual. 'Good food is only to be found in the Legion,' he says, holding out a haversack containing the regulation ration: a hunk of bread, dried sausage, a flagon of 'red' and a mug. While 'Le Boudin' is played the officer does his duty by the flagon. According to the rules of the game it contains an explosive mixture of spirits, an astounding, if not noisome, cocktail containing aniseed, gin and sweet wine. 'Thus,' explains the senior subaltern, 'our parting will be less grievous.'

Sometimes, as a final tribute to his men, the departing officer will throw down from the top of the gangway, like a bouquet of flowers, his white *képi* with its grenade badge.

The story of the six-year war in Algeria cannot be told in detail. It was a long-drawn-out process, producing few major battles and not many medals. Nor was it conducted solely by the Legion, though the Legion played a worthy part – its casualties totalled 63 officers, 264 NCOs and 1628 legionaries.

Some of the Legion were luckier; a few even found a chance of acting the hero and 'doing a Camerone'. This chapter cannot close without recounting the last of these battles when the odds were a hundred to one.

It was January 1960, and the 2nd Foreign Legion had just celebrated Christmas. It was operating in the Ain Sefra area on the edge of the Sahara.

A message arrived from Colonel Romet, the regimental commander: 'A rebel *katiba* has crossed the western barrier. You are ordered to pursue it.'

The 2nd Foreign Legion set off on the trail. Next morning

Colonel Romet was told that the enemy had been located in the Beni Smir djebel. 'About six hundred men, apparently well entrenched.'

'How are they armed?'

'They've got new up-to-date equipment.'

Captain Grosjean's company was detailed to embark as the leading unit in the transport helicopters. The idea was to land near the enemy unit and hold it down until the arrival of reinforcements which were converging in the djebel.

'Target in sight!' The leading helicopter, an H34, was poised above Hill 1641, a rounded tump swept by the wind and covered by sickly alfalfa bushes.

'I am going down,' H34 announced. It stopped three feet from the ground. The despatcher signalled to Sergeant Sanchez-Iglesias' group of a corporal and four legionaries to jump. Then the helicopter took off with a deafening roar.

'Damn!' said Galejski, the corporal. 'Where are the others?'

Sanchez-Iglesias surveyed the horizon. Nothing. The H34 had put them down in the wrong place. 'A fine pickle!' he went on. 'Here we are all alone on this ridiculous tump.'

He never finished his sentence. From all directions came a hellish fire. By bad luck the helicopter had landed the six men precisely in the middle of the rebel force. For security reasons the other helicopters did not land. The legionaries were only thirty yards away from the Fellagha.

'We must hold out,' Sanchez said. The defence was organised. From the air Piper Cubs and T6s pinned the enemy to the ground – for a time.

Suddenly a voice was heard: 'Legionaries, Morocco is only a few miles off, less than an hour's march.' The six men were being offered a choice – escort to safety in Morocco. There they could ask for political asylum, return to their native country and live in peace. If they refused, it would be massacre – without mercy.

The legionaries replied: 'Come out into the open.' Throughout the morning they remained flat on the ground, fighting at odds of twenty to one.

Suddenly, through the din of the T6s, could be heard the noise of powerful engines. Reinforcements! Their comrades had not left them in the lurch. The tall grass was flattened by the blast of the rotor; it was an H34 – a big helicoptor.

An officer, Lieutenant Cardonne, jumped down but at once the helicopter was struck by a storm of lead. The pilot opened up his engines and rose without warning. The officer, still in his harness, was carried up with it and dragged back into the machine by his men. 'Reinforcements are not coming just now,' Sanchez observed. The six men were alone once more. They had to go on holding out.

'Get a move on!' Ever since the morning Major Kopf had been hurrying along his three companies and the commando of which he was in charge. Every man was determined to reach those isolated few still fighting hand-to-hand. Major Kopf's force moved from rock to rock and cover to cover, firing all the time. Each time they passed overhead the T6 pilots reported over the radio: 'Bambi to ground troops. Your chaps are still holding. Hats off to the Legion!'

Dusk was falling. The six men had now been fighting for ten hours. Night would mean the end of air support. The rebels knew it; in the darkness they would be able to cross back into Morocco. Accordingly they redoubled their efforts against this thorn in their flesh. They had to finish with it by sundown.

Nevertheless the fighting continued. 'Surrender' came the call. Sergeant Sanchez' little band remained unmoved, ready to do its duty to the end. It was right. 'Listen,' Corporal Galejski shouted. Firing could be heard to the south. Their friends were there. The three companies and the commando arrived at last and joined up. The isolated few were relieved.

After a short rest operations continued. A helicopter pilot was thunderstruck – 'What? Those fellows had held out for ten hours against odds of twenty to one and then they calmly fall back into their place as if they had done nothing remarkable! But they've done a Camerone!'

Sergeant Sanchez-Iglesias, Corporal Galejski and legionaries Gerlich, Hortzkow, Dahmen and Paumart merely replied: 'We did what we've been taught to do in such a case and we did it efficiently. It's just our job to hold out to the end.'

Anne-Marie

September 1962

SERGEANT-MAJOR FRANKE passed his only hand over the stump of his left fore-arm. It was a well-known gesture, almost involuntary. He had just completed his 'fifteen years, six months', the regulation period giving him the right to retire. To complete the formalities he had returned to 'headquarters', the Mother Convent – Sidi-bel-Abbès.

Fifteen and a half years' service in the Legion had given Sergeant-Major Franke a thin angular face with bony cheeks and sunken eyes, at the corners of which were crow's-feet betraying his age. Franke admitted to thirty-five but in fact he was ten years older. This was no vain desire to remain young; he refused to become one of the 'old ones'.

In general terms Franke was considered a 'card'; during an evening beat-up he would calmly draw back his sleeve to bare the stump of his arm and move in to the attack. It was a weapon as formidable as a pick-helve; having laid his victims out flat, he would show them his great blue tattoo mark – three letters: MMF. '*Moi, m'en fous*' ['Don't care a damn'], he would explain with a burst of laughter.

'Don't care a damn.' This was the phrase with which, for fifteen years, Franke had met all the vicissitudes of military life – Dien Bien Phu? – MMF; the 'clink'? – MMF; the djebels? – MMF. By this time it had become more a philosophy than a profession of faith.

'You can say "To hell with it all" now,' the warrant officer of No. 2 Transit Company, the company responsible for men on release, said to him. The war was over. The Legion was going back to France. Tomorrow, or in a week at most, it would be established in Aubagne.

Franke shrugged. He glanced out of the window at the Legion's War Memorial which legionaries of the Training Company were in process of dismantling. Yes, the Legion was moving house but that was no reason to stop training recruits. The Legion would build itself a barracks elsewhere and it was taking its trophies with it, its traditions too – its sole heritage.

'To think that one is going to leave all that to "them".' Involuntarily Franke had said out loud what everyone here was silently thinking. Officially the war had ended ten weeks ago and Algeria was now independent. In Sidi-bel-Abbès the 'Fellouzes' pushed people off the pavements; they gave themselves airs as victors, even though, in their hearts of hearts, they did not really know what they had done to become so. In Sidi-bel-Abbès, as elsewhere, they paraded about, chests thrown out, tommy guns ostentatiously at the ready, in their camouflage uniform with a silk scarf round the neck, a beltful of grenades and peaked cap pulled down over the ears. They patrolled through the town, proud and swaggering.

Then, suddenly, at a crossroads, they met a patrol of the Legion. The Arabs jeered, trying to play the victor. It did them no good. The legionaries passed on impassively, chins held high, looking straight to their front, as if they had seen no one. The headquarters of the Algerian Army would have liked the right to patrol the town to be restricted to its own units. It talked of provocation. The Legion took no notice.

Equally the Legion ignored the orders from French headquarters to remain in barracks. It was a hallowed tradition: whether ten, twenty or even alone, legionaries leaving the corps had the right, as they had had yesterday or twenty years ago, for ever in fact, to be played out by the band. So, in Sidi-bel-Abbès station, beflagged with the green-white-red colours of Algeria, it was nothing unusual to hear 'Le Boudin' drowning the thin Arab music several times a day. It sounded all the louder in the empty streets.

In July the FLN had held its Victory Parade. That day the Legion had kept its gates closed and had ordered a vast barrack-cleaning fatigue with a weapons inspection on the barrack square. Throughout the morning, all day in fact, interminable files of 'Fellouzes' had marched, galloped or shuffled in front of the gate of Viénot barracks. They danced, they shouted, they fired volleys. They hoped at least to arouse the legionaries' curiosity. Not a man lifted his head. They did not belong to the same world.

Next day the FLN collected all its available troops from the whole area. It was determined to organise a 'real' parade to display its strength to the legionaries. 'It was a total fiasco,' Franke says. It would have been ludicrous, had it not been true that it was the 'Fellouzes' who now pushed people off the pavements. To the legionaries, however, remained their pride – pride of the bravest soldiers in the world, depositories of the tradition of a thousand skirmishes, a hundred battles, twenty campaigns. The American Marines, who are sparing of praise, had said that the Legion was a worthy ally. 'MMF,' Franke said with his habitual calm.

The previous day, on the 'legionaries' square' and at the town cemetery, the 700 recruits of the Training Battalion together with the NCOs and 'transitory customers' such as Franke, had paid their last stoic tribute to their dead comrades. Only three of them were to be taken back to France. three great men who had served to the death with 'Loyalty and Honour'. They included General Rollet, the father of the Legion.

But this was no moment for pathos or regrets. Franke was waiting for his papers. 'Hell,' he said to the sergeant, 'can't you get a move on.' 'Don't get excited,' the other replied. 'You'll leave on time. So shall we for that matter.'

Franke shrugged. Papers or not, for him the Legion was finished. He repeated the word – 'Finished.'

'We don't usually have as long a record of service as this,' the sergeant went on, bending over his typewriter. 'Fifteen years in the Legion doesn't mean fifteen years in the office.'

Franke smiled. An office! He had only been in one twice – in 1947 when signing on and now today. Between these two times he had made war.

In 1947 he had been sent to Madagascar; then had come a few months' tour of duty in Indo-China – a rebellion to suppress – just another one and for no good purpose; in 1948 Indo-China again for another rebellion and in 1951 Indo-China yet again – the same rebellion which had never stopped. 'I shall have spent my life hunting rebels.' Then, when Indo-China was finished came Algeria for six years – primary target, rebels yet again. 'That lot – you can't say they aren't persistent.'

Franke stowed away his papers. He was now ready. Tomorrow he would go with the Legion convoy to Oran. Then back in Europe those fifteen years would slip into the background as they had slipped away since he joined the Corps fifteen years ago. He still had his left hand then and his illusions which had survived even a two-year incarceration behind Soviet barbed wire.

He remembered the reception formalities at Fort Saint-Jean – office after office – the hospital – enrolment when an officer had asked him his identity, his background and his origin – medical examination. It was all a complex procedure through which, over one hundred and thirty years, the Legion had passed 340,000 men. So it was well run in.

Then had come his first crossing of the Mediterranean. He had admired the sunsets, the roadstead of Oran and the railway from Oran to Sidi-bel-Abbès. Like a good German he had appreciated the quality of the band and of the singing. Straight away he too had sung with his mates. There was 'Westerwald' which the parachutists had made their own, and then 'Anne-Marie', sung more slowly and more solemnly by the Legion, as if it was a profession of faith:

Mein Regiment, Mein Heimatland – My regiment is my home.

Franke was in the NCOs' mess. He was just ruminating and

for a moment forgot that he didn't care a damn. He felt like an orphan, already excluded from the family. Fifteen years – some of his mates dead, others gone. Hard times – peculiar times.

'Tell me, Franke, do you remember?'

Franke remembered everything – opening the road between Cao Bang and Bac Kan, the ambush on Route Coloniale 5 when Major Cabaribère was commanding, mines at Bac-ninh and Khenchela, the hills of Dong Khe and Ouarsenis, the mists of Laos and the snow of the Aurès mountains, fire-crackers in Cholon and beat-ups in Bougie, Camerone celebrations in every conceivable climate, even on a cruise liner, Christmas in Djibouti with the telegraph poles dressed up as Christmas trees and real Christmases at Sidi-bel-Abbès.

'You're right. I remember.'

Franke was drinking. He was not drunk. He would never be able to get drunk again because a real drinking match only takes place between buddies – within the family. Night had long since fallen. On the square of Sidi-bel-Abbès a few scattered stones were all that remained of the War Memorial. With all its traditions it had been removed that afternoon, like everything else.

'It's almost as if it had never existed,' Franke tried to tell himself.

Outside the seven hundred recruits of the Training Battalion stood in silence. The Legion's museum had gone and with it Juarez' sabre, Abd el Kader's muskets and the rebel banners. Captain Danjou's hand in its glass case had been removed with religious care. The marble plaques carrying the names of the Legion's dead had been dismantled. Nothing was left.

Yes, something was left – some ashes. On return from Tonkin in 1885 Captain de Borelli, hero of the siege of Tuyen Quang and one of its few survivors, had presented to the museum the great silken banner captured from the Black Flags. He had made a condition: 'This trophy shall never leave Sidi-bel-Abbès. If the Legion finally goes, it must be burnt.' So, that evening, the Legion burnt the Black Flags' banner – a curious ceremony devoted to a Chinese flag! Around the little yellowish flame seven legionaries stood to attention. When the little flame finally died away it was as if the soul of the Legion had been extinguished.

Then, simultaneously, 700 torches were lighted, as if to demonstrate the Legion's will to live. In the darkness rose a song, the song which legionaries had sung for 130 years from Mexico to Papeete – 'Le Boudin'.

The torches were extinguished but the legionaries went on singing. Franke had joined them. For that evening, just for that evening, he wanted still to be part of the family:

Mein Regiment, mein Heimatland
Mein Mutter hab' ich nie gekannt
Mein Vater starb schon früh im Feld
Ich bin allein auf dieser Welt.

My regiment is my home
My mother I have never known
My father fell early on in the field
I am alone in the world.

Franke sang but he was thinking: 'Sidi-bel-Abbès, it's finished. Finished, I tell you.' As he sang the last line – 'I am alone . . .' – his voice broke.

For Franke indeed it was finished. But today the Legion still goes on, under another sky and with other duties. It is still present.

Appendices

Appendix I

The Legion's Campaigns

War or campaign	*Battles*	*Units involved*	*Strength**	*Killed* *Offrs*	*Killed* *NCOs*	*Killed* *Men*
Algeria 1831–2	Sidi Chabel 11 Nov. 1832	Old Legion 1831–3	Kept up	27	61	756
	Moulay-is-Mail 26 June 1835	New Legion 1836–7				
	La Macta 28 June 1835					
	Zibans 1844	1st Legion Regiment 1837–55	5000 (average)			
	Zaatcha 1849	2nd Legion Regiment 1837–55				
	Takitount 1858	2nd Foreign Legion 1855–9				
Spain 1835–9	Terapegui 26 April 1836	Old Legion under Colonels Bernelle, Lebeau & Conrad	3000 (maintained)	28	98	977
	Huesca 24 March 1837					
	Barbastro 2 July 1838					

* Figures given are those permanently maintained up to the level of paper strength.

War or campaign	Battles	Units involved	Strength*	Killed Offrs	Killed NCOs	Killed Men
Crimea 1854–5	Alma, Sebastopol June 1855	1st & 2nd Regiments, 2nd Foreign Legion	4500	25	32	387
Italy 1859	Magenta, Solferino June 1859	1st & 2nd Foreign Regiments	4500	4	11	128
Mexico 1863–7	Camerone 30 April 1863	Foreign Legion Regiment	3000	22	32	414
France 1870–1	Orleans 11 Oct. 1870 Montbéliard	Foreign Legion Regiment (three battalions)	3000	14	52	864
Southern Oran 1882–1907	El Mungar 2 Sept. 1903	1st & 2nd Regiments Mounted companies	Variable	8	46	601
Tonkin	Sontay, Tuyen Quang 1883–1910	Two battalions (in formation)	3000 (maintained)	23	159	1882
Formosa 1885			Token detachment	3	6	24
Dahomey 1892–4	Dogba 19 Sept. 1892	1st Foreign Battalion (mobile)	600	2	4	31
Sudan 1893–4			Detachment			2
Madagascar 1895–1901	Tananarive 1898	1st Battalion, 1st Foreign Legion	1200 (maintained)	5	27	228

War or campaign	*Battles*	*Units involved*	*Strength**	*Killed*		
				Offrs	*NCOs*	*Men*
Morocco 1907–14	Ouled Soltane 1912	2nd Regiment	3000 (maintained)	5	21	299
France 1914–18	Somme 1916 Aisne 1917 Verdun 1917 Hindenburg Line 1918	Four foreign battalions to 11 Nov. 1915. Then Mobile Foreign Legion Regiment	13,000 (maintained)	139	349	3628
Middle East 1914–18	Gallipoli 1916 Dardanelles 1917	1st Middle East Battalion	1600	16	78	721
Morocco 1914–18		1st Legion Regiment		4	40	304
Tonkin 1914–40		Four battalions (in formation). Then 5th Foreign Legion	3000 (average)	1	5	49
Morocco 1920–35	Taza 1933 Djebel Baddou 1933	2nd Foreign Legion 3rd Foreign Legion 4th Foreign Legion Mounted companies		74	158	1264

War or compaign	*Battles*	*Units involved*	*Strength**	*Killed* *Offrs*	*NCOs*	*Men*
Syria 1925–7	Messifré 16 Sept. 1925	1st Foreign Legion Cavalry 5th Battalion, 4th Foreign Legion	3000	2	6	37
France 1939–45	France May 1940 Narvik June 1940 Bir Hakeim 1942 Tunisia 1943 Italy 1943 France 1944	11th Foreign Legion 12th Foreign Legion 13th DBLE 13th DBLE, plus mobile regiment Foreign Legion		118	821	8078
Indo-China 1945–54	Phu Tong Hoa 1948 Langson 1950 Dien Bien Phu 1954, etc	2nd Foreign Legion 3rd Foreign Legion 5th Foreign Legion 13th DBLE 1st Foreign Legion Cavalry 1st & 2nd Foreign Legion Parachute Regiments	30,000 (permanently)	309	1082	9092

War or campaign	Battles	Units involved	Strength*	Killed Offrs	Killed NCOs	Killed Men
Madagascar 1947–50		1st Regiment		3	1	1
Tunisia 1952–4		2nd Foreign Legion	1500	2	1	11
Morocco 1953–6		4th Foreign Legion, plus Sahara companies		3	7	56
Algeria 1954–61 (to 1 March 1961)		1st, 2nd, 3rd, 4th, 5th Foreign Legion 13th DBLE 1st & 2nd Foreign Legion Cavalry 1st & 2nd Parachute Regiments Sahara companies	20,000 (permanently)	63	264	1528

Appendix 2

History of the Legion

I. *Organisation*

Mar. 1831–Dec. 1840	Foreign Legion
Jan. 1841–Dec. 1854	1st & 2nd Regiments of Foreign Legion
Jan. 1855–Apr. 1856	1st & 2nd Foreign Legion
Apr. 1856–Dec. 1861	1st & 2nd Foreign Regiments
Dec. 1861–Mar. 1875	Foreign Regiment
Mar. 1875–Dec. 1884	Foreign Legion
Dec. 1884–Nov. 1920	1st & 2nd Foreign Regiments
1910–20 in eastern Morocco	1st Mobile Regiment (Régiment de Marche) of 1st Foreign Legion
1910–20 in western Morocco	1st Mobile Regiment of 2nd Foreign Legion
1914–15 in France	2nd Mobile Regiment of 1st Foreign Legion
	3rd Mobile Regiment of 1st Foreign Legion
	4th Mobile Regiment of 1st Foreign Legion
	2nd Mobile Regiment of 2nd Foreign Legion
	(These four regiments combined to form the Foreign Legion Mobile Regiment)
1915–20 in France	Foreign Legion Mobile Regiment
Nov. 1920	1st Foreign Regiment (combined depot in Algeria)
	2nd Foreign Regiment (Morocco)
	3rd Foreign Regiment (direct successor to Foreign Legion Mobile Regiment – Morocco)
	4th Foreign Regiment (Morocco)
1922	1st Foreign Legion Cavalry

II. *Units after 1920*

1 July 1945	Mobile Legion Infantry Regiment (in Far East)
	converted into
1 Jan. 1946	2nd Foreign Legion Infantry
1 Jan. 1921	3rd Foreign Legion Infantry (Mobile Regiment reconstituted)
17 Dec. 1920	4th Foreign Legion Infantry – the Legion's Moroccan battalion (disbanded 9 Nov. 1940)
23 Mar. 1943	1st Foreign Legion Mobile Infantry
1946	4th Half-Brigade of Foreign Legion (Morocco)
1 June 1951	1st Battalion of 4th Foreign Legion Infantry (independent unit)
1 Apr. 1955	4th Foreign Legion Infantry
1 Sept. 1930	5th Foreign Legion Infantry
2 Sept. 1939	6th Foreign Legion Infantry (independent unit)
2 Oct. 1939	6th Foreign Legion Infantry (date of formation)
1 April 1949	6th Foreign Legion Infantry
Oct. 1939	11th Foreign Legion Infantry
Feb. 1940	12th Foreign Legion Infantry
	21st Mobile Regiment of foreign volunteers
	22nd Mobile Regiment of foreign volunteers
	23rd Mobile Regiment of foreign volunteers
Feb. 1940	13th Half-Brigade of Foreign Legion
1 July 1939	2nd Foreign Legion Cavalry
Feb. 1940	97th Divisional Reconnaissance Group
Tunisia 1943	Independent Foreign Legion Cavalry Group
1 July 1948	1st Foreign Legion Parachute Battalion
1 Nov. 1948	2nd & 3rd Foreign Legion Parachute Battalions
1 Aug. 1955	1st Foreign Legion Parachute Regiment
1 Dec. 1955	2nd Foreign Legion Parachute Regiment

Appendix 3

Administrative and Recruiting Regulations

1. *Conditions of Enlistment*

Be physically fit for service in the armed forces and capable of active service in any overseas theatre of operations.

Minimum age 18 years, maximum age 40 years.

Minimum height 5 ft 1 in.

2. *Terms of Enlistment*

The period of enlistment is five years.

Re-engagements may be for periods of six months to five years.

3. *Identity Papers*

These are not required. Candidates may enlist under their real identity or a fictitious one.

4. *Where to Report*

On French territory candidates for enlistment are accepted by the following agencies:

Offices of Military Administration
Gendarmerie headquarters
Garrison headquarters
Military units

and in the Foreign Legion's information centres at Vincennes, Marseille and Strasbourg.

On arrival at a centre and after examination the candidate will sign a provisional enlistment contract.

After a further examination the contract will be definitively ratified at the Foreign Legion depot. It takes effect from the day of signature of the provisional contract.

5. *Enlistment Bounty*

The enlistment bounty is 1320 francs.

A supplementary bounty is subsequently granted to legionaries who hold an easily obtainable certificate.

6. *Pay*

The pay scale is that applicable to the French army.

7. *Retirement Pension*

The graduated retirement pension is due after fifteen years' service.

The seniority retirement pension is due after twenty-five years' service.

For other ranks the retirement pension is calculated on the basis of a sergeant's annual pay.

8. *Promotion*

Dependent on their capabilities promotion to any rank is open to all legionaries.

9. *Adjustment of Civil Status*

NCOs and legionaries who, on enlistment, declare a civil status which is not juridically valid, may be permitted, after a certain period, to resume their genuine civil status.

10. *Naturalisation*

After adjustment of their civil status legionaries of foreign origin may request naturalisation under the normal conditions.

The quality of the legionary or ex-legionary and primarily his record of service may noticeably accelerate the procedure.

Appendix 4

The Official Account of Camerone

(as read on 30 April each year to all troops of the Foreign Legion)

The French army was besieging Puebla.

The Legion had been ordered to protect traffic and ensure the security of convoys over seventy-five kilometres of road.

On 29 April 1863 the commanding officer, Colonel Jeanningros, was informed that a large convoy, carrying three million francs in specie, siege material and ammunition, was *en route* for Puebla. Captain Danjou, his second-in-command, persuaded him to send one company ahead of the convoy. No. 3 Company of the Foreign Legion Regiment was detailed but it had no officers available. Captain Danjou took command himself and 2nd-Lieutenants Maudet of the colour party and Vilain, the paymaster, volunteered to join him.

At 1 a.m. on 30 April No. 3 Company set off at a strength of three officers and sixty-two men. By 7 a.m. it had covered some twelve miles when it stopped to brew coffee at Palo Verde. At this moment the enemy appeared and fighting started at once. Captain Danjou formed square and, during his withdrawal, successfully repulsed several cavalry charges, inflicting his first first severe losses on the enemy.

On reaching Camerone inn, a vast building including a courtyard surrounded by a 9-ft-high wall, he decided to dig in there in order to pin the enemy down and delay to the maximum the time at which the convoy might be attacked.

While the men were hurriedly organising the defence of this building a Mexican officer summoned Captain Danjou to surrender, pointing out the vast disparity of numbers. Captain Danjou replied: 'We have ammunition and we shall not surrender.' Then, raising his hand, he swore to defend himself to the death and required his men to swear the same oath. It was now 10 a.m. Until 6 p.m. these sixty men, who had neither eaten nor drunk since the previous day and despite extreme heat, hunger and thirst, defended themselves against 2000 Mexicans – 800 cavalry, 1200 infantry.

At midday Captain Danjou was killed by a bullet through the heart. At 2 p.m. 2nd-Lieutenant Vilain fell with a bullet in the head. At this point the Mexican colonel succeeded in setting fire to the inn.

Despite their sufferings, increased by heat and smoke, the legionaries held on but many of them had been hit. By 5 p.m. there remained only twelve men fit to fight, grouped around 2nd-Lieutenant Maudet.

At this point the Mexican colonel assembled his men and emphasised to them the disgrace which would attach to them if they were unable to overcome this handful of brave men (a legionary who understood Spanish translated his speech as it was made). The Mexicans prepared for a general assault through the breaches which they had succeeded in opening but Colonel Milan first sent another surrender call to 2nd-Lieutenant Maudet. It was scornfully rejected.

The final assault took place. Soon there remained only five men round Maudet – Corporal Maine and legionaries Catteau, Wenzel, Constantin and Leonhart. Each still had one round in their rifle and their bayonets were fixed; in a corner of the courtyard, their backs to the wall, they faced their enemies; at a given signal they fired point-blank at the enemy and then charged with the bayonet. Second-Lieutenant Maudet and two legionaries fell, mortally wounded. Maine and his two legionaries were about to be massacred when a Mexican officer dashed forward and saved them, shouting 'Surrender'. 'We will surrender if you promise to collect and care for our wounded and if you leave us our weapons,' Maine replied. The three men still had their bayonets lowered. 'We can refuse nothing to men like you,' the officer answered.

Captain Danjou's sixty men had kept their oath to the end; for eleven hours they had held up 2000 enemy, of whom they had killed 300 and wounded a similar number. By their sacrifice and by saving the convoy, they had carried out their orders.

The Emperor Napoleon III decided that Camerone be inscribed as a battle honour on the colours of the Foreign Legion and that the names of Danjou, Vilain and Maudet be engraved in letters of gold on the walls of Les Invalides in Paris.

In 1892 a memorial was built on the scene of the fighting.

Ever since then, when Mexican troops pass the memorial, they present arms.

Bibliography

This bibliography is necessarily incomplete. It includes only those books described as 'essential' by the Foreign Legion's Historical Service, and other books used by the author.

PUBLISHED IN FRANCE

Aage, Prince of Denmark: *Mes Souvenirs sur la Légion Étrangère; Nouveaux Souvenirs sur la Légion Étrangère*, Payot, Paris 1936.

Azam, Captain H.: *La Légion Étrangère, ses règles particulières*, Presse de Képi Blanc, Sidi bel Abbes 1951; *Pages de gloire de la Légion Étrangère*, Martel, Paris 1952.

Blond, Georges: *La Légion Étrangère*, Stock, Paris 1964.

Bonnecarrère, Paul: *Par le sang versé*, Fayard, Paris 1969.

Fieffé, Eugène: *Histoire des troupes étrangères au service de la France*, Librairie Dumaine, Paris 1854.

Képi Blanc: Monthly Journal of the Foreign Legion, SMOLE Aubagne 13.

Le Livre d'or de la Légion étrangère (1831–1931), Imprimerie Frazier Soye, Paris 1931.

Le Livre d'or de la Légion étrangère, Charles Lavauzelle & Cie, Limoges 1958.

MacOrlan, Pierre: *La Légion étrangère*, Flammarion, Paris 1933.

Maire, Colonel: *Souvenirs de Colonel Maire sur la Légion étrangère*, J.-P. Dorian, Paris 1943.

Manue, Georges: *Têtes brûlées. Cinq ans à la Légion*, Societé d'Édition, Paris 1929.

Mordal, Jacques: *Bir Hakeim*, Press Pocket, Paris 1970.

Poirmeur, Commandant H.: *Notre vieille Légion*, Berger-Levrault, Paris 1931.

Raulet, André: *Légion über alles! Souvenirs sur la Légion étrangère*, Charles Lavauzelle, Paris 1934.

Vallières, Jean des: *Sous le drapeau de la Légion étrangère. Les Hommes sans nom*, Albin Michel, Paris 1933; *Sous le drapeau de la Légion étrangère, Sa grandeur: l'infortune*, Albin Michel, Paris 1945.

PUBLISHED IN GERMANY

Bamberg, G.: *Aus Afrika Spanien-Iena*, Fr Manke, Berlin 1870.

Rahden, Wilhelm von: *Wanderungen eines alten Soldat. Aus Spaniens Bürgerkrieg 1833–1840*, Verlag D.G. Ober-Hofbücherei, Berlin 1851.

PUBLISHED IN SWITZERLAND

Algier, Mexico, Rom. Schiksale eines Schweizers im fremden Kriegsdienst, Schroeter, Zürich 1901.

Augustin, Le: *Combattant volontaire suisse au régiment de marche de la Légion étrangère*, Imprimerie Centrale, Lausanne 1934.

Plessis, Theodore de: *Souvenirs de mes campagnes à la région étrangère (Mexique et Afrique 1865–1868)*, A. Julien, Geneva 1923.

Spinner, Henrich: *Les Souvenirs d'un vieux soldat* (Part 1: *A la Légion au Mexique*; Part 2: *Dans les troupes papales*), Imprimerie Messelier, Neuchatel 1906.

PUBLISHED IN BELGIUM

Lamborelle, Louis: *Cinq ans en Afrique*, Brussels 1863.

Macquet: *Correspondance d'un soldat belge en Crimée*, Flamme, Tournai 1858.

Notice sur la Légion étrangère, Rouvroy, Namur 1843.

PUBLISHED IN LUXEMBURG

Le livre d'or de nos légionnaires (Patria 1914–1918), Imprimerie Centrale Gustave Soupert, Luxemburg 1930.

PUBLISHED IN POLAND

Polonia – Polish weekly review, issue of 21 July 1927, article entitled *Roman Ilnickt.*

PUBLISHED IN DENMARK

Aage, Prinds: *Tre Aars Kampe i Fremmed legionen*, Copenhagen 1927.

Bruun, Daniel: *Ungs Domsfoerd Rundt Kap gode Haab. Feltlir i Sahara Erindringer og Dagsbogkitzer*, Copenhagen 1936.

PUBLISHED IN SWEDEN

Sternjtrom, Edward: *Minen af en Svensk Legionar*, Stockholm, undated.

PUBLISHED IN GREAT BRITAIN

Gibbons, John: *The Truth about the Legion*, Methuen, London 1933.

O'Ballance, E.: *The Story of the French Foreign Legion*, Faber & Faber, London 1961.

Seeger, Alan: *Letters, Diary, Poems*, Constable, London 1917.

Ward Price, G.: *In Morocco with the Legion*, Jarrolds, London 1934.

PUBLISHED IN THE UNITED STATES

Bowe, John: *Kelly of the Foreign Legion*, Minneapolis 1918.

Doty, G.: *The Legion of the Damned*, Century Co., New York and London 1928.

Rockwell, Kiffin Yates: *Foreign legionnaire and aviator* (War letters, France 1914–18), New York.

Rockwell, Paul Ayres: *American Fighters in the Foreign Legion 1914–1918*, Houghton Mifflin, Boston and New York 1930.

War

Readiness at Dawn	Wing Commander Ronald Adam	50p
We Rendezvous at Ten	Wing Commander Ronald Adam	50p
Operation Mercury	M. G. Comeau	40p
The Raider Kormoran	Capt Theodore Detmers	45p
Sea Eagle	E. P. Hoyt	50p
The Elusive Seagull	E. P. Hoyt	60p
Rear Gunner	John Beede	60p
Assault from Within	Georg von Konrat	60p
Pocket Battleship	Theodore Krancke & H. J. Brennecke	50p
U-Boat 977	Heinz Schaeffer	45p
The Warring Seas	A. V. Sellwood	55p
Commando	Brigadier Durnford-Slater	45p
G Patrol	Michael Crichton-Stuart	45p
RAF Biggin Hill	Graham Wallace	50p
Long Run to Tobruk	Gordon Landsborough	45p
Battery from Hellfire	Gordon Landsborough	40p
Patrol to Benghazi	Gordon Landsborough	40p
The Glasshouse Gang	Gordon Landsborough	50p
Desert Marauders	Gordon Landsborough	50p
Benghazi Breakout	Gordon Landsborough	60p
The Dead Commando	Gordon Landsborough	60p